OPEN THE HANGAR DOORS
And the Stories Fly Out

GEORGE M. LEUTE, JR
Captain USN & USNR (Retired)

OPEN THE HANGAR DOORS
And the Stories Fly Out

GEORGE M. LEUTE, JR
Captain USN & USNR (Retired)

Wallingford, PA 19086
An Imprint of
Durán Creative Solutions, Inc.

Open The Hangar Doors And the Stories Fly Out

Copyright © 2005 by George M. Leute, Jr.

All rights reserved. No part of this book shall be reproduced, stored in a retrieval system, or transmitted by any means — electronic, mechanical, photocopying, recording, or otherwise — without written permission from the publisher, except for the inclusion of brief quotation in a review.

International Standard Book Number: 0-9771935-0-0

Library of Congress Control Number: 2005932022

Printed in the United States

First printing: August 2005

09 08 07 06 05   9 8

Cover page photographs: On the ground is the F7U Cutlass (Courtesy of Vought Aircraft Industries; flying is the F-14 Tomcat (Leute photo).

Durán Press
Wallingford, PA 19086 (USA)

DEDICATION

*This book is dedicated to
Naval Aviators and their crewmen,
present, past and future.*

Contents

Acknowledgments ix
Preface xiii
About The Author xix
Author's Note xxi
The Hangar Doors Are Open xxiii

Part One: Fuel Starvation 1
Fuel Consumption and Weather — Twin Partners of Concern 3
The Flushing Toilet — Landing With Minus 250 Pounds of Fuel 8
The Flushing Toilet — Plus Bad Weather 15
The Flushing Toilet — Plus Bad Weather and Radio Problems 21
The Flushing Toilet — Plus Bad Weather and Both
    Engines Shutdown 27

Part Two: Where There's a Will 31
Where There's A Will — The Mission Is All 33
The Simple Mission? No Way! 36
In Search of the Submarine — On a Very Black Night 45
The Agony of the Longest Night 60
Ice Floes and Helicopters — A Rescue Mission 69

Part Three: Human Factors 81
Human Factors — Stop the Damn Nonsense 83
Why Do They Do It? They Know Better! 85
Was This Trip Necessary? 95
    More About Crosswinds 108
    The Effect of Winds on Landing Aircraft 110
What Was "Poor" Chuck's Problem? 112
The Judge's Embarrassing Mistake! 119

Part Four: Emergency Bailouts 129
Emergency Bailouts — High, Low, and Other 131
High-Altitude Bailout: Not a Piece of Cake 134
Low-Altitude Bailout: Not a Piece of Cake 144

Introduction   vii

Bailouts on the Deck and Underwater: It's Hairy 155
Self-Induced Bailouts: No Piece of Cake Either 169

Part Five: Formation Flight **177**
Formation Flight — A Means to an End! 179
The Air Group Goes Aboard Ship: Patience Required 183
The Carrier Returns from Sea: Launch the Airgroup 193
Superchargers and Formation Flight: A Problem 200
Weather Plays a Cruel Trick on the VIP's 205
Swept-Wing Jets Go Aboard the Carrier: A Last-minute Change 210

Part Six: Instrument Flight **219**
Instrument Flight — Essential to Life 221
    Simple Principles of Lift 224
    The Artificial Horizon 226
    The Life Saver — The Turn and Bank Indicator 228
Instrument Flying at Its Best: The "Pure Fun" of It! 234
NAS South Weymouth, MA, to NAS Willow Grove, PA 249
The Distractions of Smog and Catalina Island Near Los Angeles ... 256
An Instrument Flight Puzzle? 260

Part Seven: Mechanical Problems **265**
Mechanical Problems — Surprises and Choices 267
Engine Regulators Gone Amuck 272
The Steady, Reliable Avenger 279
The Pros Don't Quit 285
A Life or Death Situation — What Would You Do? 290

Part Eight: More Surprises Bring More Challenges **299**
More Surprises Bring More Challenges 301
A Fine Mess We've Gotten Into! 307
Things You Just Have To Respect 314
    Believe! Your Life May Depend Upon It 322
A Flight Worthy of a Distinguished Flying Cross 325

Epilogue **343**

ACKNOWLEDGMENTS

It is in the tradition of telling stories about Naval Aviation and Naval Aviators that they be described orally, with one story suggesting another and that story suggesting another story in an almost endless stream. Each aviator listens but also waits for the chance to contribute. These stories flow out continuously because they are interesting; and they are interesting because they are unusual or exciting — often too exciting — and they are diverse, in most cases evolving out of that aviator's experience or what he knows of some other aviator's experiences.

Telling about these events is one thing; writing about them in such a way as to interest those with military flight experience and, at the same time, readers with little knowledge about the demands of aviation is more challenging. With this in mind, it would have been easy to be discouraged and decide that the result would not be worth the effort or that the result would not be worthy of the pilots whose actions are narrated here. So, a natural reluctance seemed to predominate, a reluctance which my wife Eileen persuaded me to overcome.

But there are others to whom I owe thanks. Originally, the idea was to include only those experiences that Naval Aviators had had that never appeared in any kind of publication; that is, they were close calls which were nearly always successfully resolved or they were incidents, often embarrassing, which the pilots created themselves, but which they resolved, leaving most people unaware or their occurrences. These incidents remain the main part of the book. There were, however, a few incidents which appeared in various publications, which are partially or wholly included here. These incidents seemed so significant for a well-rounded presentation, both

Introduction

to include representative periods of time and to include representative aircraft of those periods, that their omission would have been an oversight. Referred to here are experiences of Naval Aviators like former Blue Angels leader CAPT Ray Hawkins USN, former test pilot and astronaut CDR Joe Edwards USN, CDR Russ Pearson USN, LCDR Jim Qualls USN, and LT Francis Lagan USNR. Their experiences are described in detail.

For information on some of these, I am indebted to The F-14 Tomcat Association, The Tailhook Association, and to CAPT David L. Parsons, USN, a former editor of *Approach* magazine, the Navy's official flight safety magazine. CAPT Parsons, who has written a book *Fighter Country*, was kind enough to offer help in research, if needed; and although considerations of space and the desire to include representations of various types of aircraft and incidents over longer periods of time precluded accepting his assistance, his kind offer was yet more encouragement to follow through on this book.

Another source, with which anyone familiar with Naval Aviation is acquainted, is *Naval Aviation News*, a magazine of long standing which highlights all sorts of brief news stories and cryptically describes interesting — often very exciting — incidents involving Naval Aviators and their flying travails. An ongoing feature of that magazine is Grampaw Pettibone, who is "envisioned as a cantankerous old Naval Aviator railing against present-day flyers whose unsafe actions cause aircraft accidents." His observations, whether indicating exasperation, displeasure, amazement, satisfaction, sadness, or whatever, are punctuated with insightful and understanding invective which are prefaced in old-fashioned expletives: "Great Balls of Fire;" "Leaping Lizards;" "Holy Hannah;" "Gol Dang It;" and on and on. A quote from one issue is typical: "Sufferin' catfish? This one takes the cake? I frankly must admit that this aircraft did a heck of a lot better without the pilot." Although the original Grampaw Pettibone, Robert C. Osborn, passed away in 1994, he lives on in others who have assumed his role. The importance of Osborn and his successors lies in the legacy: "Gramp's pithy remarks remain in the hearts and minds of aviators, who extol countless times when the old-timer's safety messages saved them from flight accidents." His acknowledgment here is an indication of the part he still plays when Naval Aviators get together.

And, how could I, or anyone else undertaking a book like this, not acknowledge and show respect to the other aviators, the crewmen, and the mechanics who had a part in each experience, more than a few of which are from my years of flying.

Last, and very important, is the help of my publisher, Ana Duran, whose advice, skills, and assistance were essential if this book was ever to become a reality.

PREFACE

Anyone picking up this book would naturally think that it is about aircraft, Naval Aircraft. But it is about more than just aircraft; more importantly, it is about people, people facing challenging situations. Yet even more, it is about the decisions and judgments, sometimes made instantaneously, that people (Naval Aviators) make in attempting to deal with difficult, puzzling, frequently life-threatening situations they encounter. These decisions are forced upon the pilot; he has no luxury of indecision. Wrong decisions can have tragic consequences, not only for the pilot but often also for crew members, for shipmates on the aircraft carrier, or people on the ground. Judgments and decisions are required for problem resolutions, resolutions, which, one way or the other, the Naval Aviator "has to live with" as the saying goes — or in some cases, has to die with. Additionally, this book is not intended only for Naval Aviators; they are relatively small in number, with, according to the Navy Personnel Command, a total of only 161,480 having been designated from Naval Aviation's beginning in 1911 through the year 2003, an average of about 1,755 per year.

Of course, the training of Naval Aviators has not occurred in such an orderly and even progression, many more being trained in periods of war, particularly World War II compared to a current rate of about 1,240. When one considers the number of people who have lived in those ninety or so years, or when one considers our present population of approximately two hundred ninety-five million, it is easy to put into perspective the very small number of people who have had the opportunity and experience of serving as Naval Aviators. While most of these pilots, like many young people flying in World War II, who accomplished such marvellous things in the Pacific Theater with relatively little flight time, flew for only a few years, many flew for twenty or more years. But whether a Naval Aviator has flown for many years or for only a few, each has had experiences or observed incidents: accidents, acts of skill, acts of courage, and exciting or exceptional experiences, which reveal the challenges of a seemingly infinite variety of occurrences which make up Naval Aviation. While the total of each pilot's experiences and observations would be quite different from every other's, depending on the types of aircraft flown, length of service, and particular years of service, practically everything has been witnessed by many others, testifying to the facts of each situation. Thus, should every Naval Aviator commit his experiences to paper, nearly every revelation would appear also in the writings of some of the other's, varying somewhat according to each one's perception. It is surprising to this writer, and I am sure to many other people as well, that so little has been done to describe these experiences for the general public.

This day of pilotless aircraft, including the Predator and the Global Hawk, versatile drones which can carry out missions or provide real-time information to computers on the ground — and satellite navigation, automatic landing equipment, long-range missiles, and who-knows-what in our age of rapidly developing technology — make some believe the pilot may be a relic of the past. The Army's modern, technically equipped and trained Fourth Armored Division, which was not fully deployed early in the war against Iraq, made some wonder whether the foot soldier is now a thing of the past; but events have largely disproved that thinking. However that may be, it is difficult to imagine bombers flying from some base in our Mid-West to targets thousands of miles away, delivering their payloads, and returning thousands of miles to their home base safely — without pilots at the controls! Yet even such things are not visionary. The pilotless Predator reconnaissance aircraft was used effectively in Afghanistan and Iraq and may be the precursor of things to come. Already well along, the X-47A Pegasus strike fighter, being readied by Northrop Grumman, is fully automated. An article in *US News and World Report* in July 2003 emphasizes its advantages over humans in being immune to G-Forces and long exhausting hours of flight time. It has the potential also of making carrier landings, provided its successful simulated carrier landing on a runway is proof enough. Yet the same article stresses that unmanned aircraft have a poor safety record and that "there will always be a need for pilots." Rather it is expected that manned and unmanned aircraft will work "in concert."

Naturally, the narratives of today's Naval Aviators flying F-14 Tomcats, F-18 Hornets, A-6E Intruders, S-3B Vikings, P-3 Orions, etcetera, would differ widely from those of the first Naval Aviators, those legendary pioneers like Ellyson, Chevalier, Mustin, Whiting, Towers, and others. But then, the pioneers' experiences flying primitive aircraft and operating from makeshift aircraft carriers, at a time of practically nonexistent electronics equipment, would provide stories no less exciting than those of today's supersonic jets, supercarriers, and unbelievable, and exceptionally reliable, electronics equipment. Many would argue that advances in technology, including systems backups or redundancy, computer-operated systems, sophisticated flight simulators for specific models, reliable generators, marvellous electronics, advances in aircraft carrier design, and satellite navigational systems actually make flying easier today than in the past. Although similarly advanced developments have been made in all of military aviation, nowhere are scientific aviation developments so obvious as in the smart, precision weapons delivered by military aircraft, which the public reads about today.

While this book could concentrate on mistakes and pilot errors, such a concentration would be skewed; it would not be fair and would distort the fine hard work, the high skills, and the acts of cool courage exhibited every day by Naval Aviators in many parts of the world and on aircraft carriers deployed in distant seas. While not concealing events which are the result of lapses in judgment or the result of faulty maintenance, this book is really intended as a tribute to the highly skilled people in Naval Aviation, who go about their daily work in a very professional manner. They have been, and remain, a valuable national asset and a source of national pride. What this book attempts to do is to look at the human side of experiences and incidents — to describe the difficult situations, and even a few that are not so difficult, which pilots have confronted and how they have reacted to them for better or worse. After all, these pilots are human beings like you or your friends or relatives. As such, they experience fear, exhilaration, and sometimes disorientation. But mostly, they control their apprehension and regain their orientation to bring about cool, rational resolutions of some very difficult, often life-threatening situations. It should be remembered that the individual military pilot does not select the weather he will fly in, nor does he select the time of day or night, nor does he select the kind of flight: the mission assigned determines all of these elements. If military aviation is to be effective, it must and does have all-weather, twenty-four-hour-a-day capability under the direction of intelligent, capable experienced leaders.

All of the events described in this book are real, and many can be verified by one printed source or another, while others not appearing in print anywhere are known only to the pilots and a few squadron mates. With but one exception, an incident at Adak, in Alaska's Aleution Islands, the author is personally aware of each incident, along with the pilot and a few others in the pilot's unit. However, the Adak incident has been described and told often enough to lend it some credence. Likewise, the oft-repeated statement that there are more aircraft accidents in routine operations than in combat had been floating around for years, yet the difficulty of finding a source or the sources for that or similar statements put into doubt the existence of responsible people who made the statements and hence the validity of such claims. Like the Adak experience, the number of times pilots had heard that idea expressed made it seem that many people in positions to know had been expressing the idea or belief on firm information based on reports of losses. This familiar statement is confirmed by one admiral, perhaps others as well. Vice Admiral Gerald E. Miller USN in speaking about the War said, "More aircraft were lost in operational accidents than in combat." And indeed, although military aircraft flying in support of the Afghanistan and Iraqi operations have faced relatively little hostile fire so far, aircraft losses from other causes still occur. While no such quotes or written descriptions about the Adak occurrence have been found by this author, the incident is included for at least two reasons: the number of times the story has been told; and how it helps to illustrate and explain the troublesome aspect of landing big airplanes on icy runways when there are strong crosswinds. What is to be emphasized is the idea that these incidents are not war stories but things that occur in the course of routine, daily flying sometimes described as "routine training flights."

Except in several instances where reference is made to an event which has appeared in a book a magazine, or other publication, the names used in these narratives are not the real names of the persons involved. But in cases in which the incident was reported in some published form, there would be no point in not using the pilot's real name, the name, for example, of someone like CDR Russ Pearson USN, who used his ejection seat while he was inverted and submerged in the Pacific Ocean, or the case of CDR Joe Edwards USN, an excellent fighter pilot, whose performance in bringing his damaged F-14 Tomcat back aboard ship while badly injured and nearly blind, gained him a Distinguished Flying Cross and led to his becoming a space shuttle pilot. There are a few others whose feats have also appeared in publications. But most of the incidents in these pages have not appeared in publications. While as a rule, it might serve some purpose to use real names of those whose actions reflect favorably on them, it would serve no purpose here to use the real names of those whose experiences would reflect unfavorably on them and thus bring unflattering attention to people whose careers are otherwise exemplary — after all, we all make mistakes, and hopefully, we all learn from our mistakes. Keep in mind that the Naval Aviator may have an exciting, even near-death, occurrence on one day or night and then get out there that very night or the next day again as if nothing had happened — though you can bet he is a little wiser. Naval Aviators involved will have no difficulty in identifying themselves in these incidents, or they may recall similar incidents, since most are either unique or else they differ in easily recognizable ways.

The reader will notice some variation in the way that incidents are related. Some are described in greater detail, including the thoughts and words of the pilots involved, while others are not. The difference is, of course, accounted for by the author's degree of friendship and knowledge of the pilots involved and also on the extent of the author's personal knowledge of the incident described. As you might expect, some of the events described are from the author's own experiences, although, in these cases also, another name is also used.

The events and incidents described in this book cover a broad space of time and include aircraft which are no longer in operation; likewise some of the pilots have long since retired and are no longer involved in Naval Aviation. Covering events which have occurred over time enables the reader to see the progress and developments which have evolved to help with the dangerous and complicated demands placed on Naval Aviators: From the Landing Signal Officer (LSO) with his "paddles" — to various experimental devices which used a light board to provide the same information as the "paddles" — to Mirror Landing Systems projecting crucial landing approach signals to the pilot — and then to the Fresnel Lens Optical Landing System — and now to an Improved Fresnel Lens Optical Landing System; parachutes and ejection seat systems have gone through amazing developments. There have been many other improvements, not to mention the change from straight-deck aircraft carriers to the angled or "canted" deck carrier, and the switch from hydraulic to steam catapults. While these have been excellent developments as far as the pilot is concerned, they have also enabled the Navy to increase its effectiveness. After all, Naval carrier aircraft are heavy, fuel-filled, high-speed vehicles.

An attempt has been made to include aircraft types representative of their missions: fighters, attack bombers, etcetera. And finally, the reader is reminded that real people are involved in all of these events. The reader is encouraged to try to place himself or herself in the situations described here — even in those very serious situations where death is or may be imminent.

About The Author

The author is a veteran Naval Aviator with twenty-four years of flying military aircraft. He has flown thirty-eight different models of Navy aircraft with concentrations of flight time in dive bombers, attack bombers, jet fighters, antisubmarine types, and long range patrol aircraft. He was among the earliest group of squadron aviators to fly the Navy's first swept-wing fighter, the F9F-6 Cougar, aboard an aircraft carrier.

Having graduated from the Navy's Instrument Flight Instructor School, he then served as an instrument flight instructor at the Naval Air Station Pensacola, Florida, and later at the Naval Air Station Corpus Christi, Texas. His first tour of sea duty was with an Attack Squadron in the first Air Group assigned to the *USS CORAL SEA* (CVB-43), an assignment which included the carrier's four-month shakedown cruise in the Caribbean. For his second tour of sea duty, he was assigned to a an experimental aircraft squadron working on carrier-related projects. A fortunate assignment, the experimental squadron exposed him to many of the new developments emerging at the time and which would be an essential part of Naval Aviation's future: the navigational system TACAN (Tactical Air Navigation System), both a land-based system and a sea-going, thus moveable, version of the Federal Aviation Agency's Omni-Directional/Distance Measuring Equipment; also, the emerging aircraft air-to-air refueling equipment and procedures; additionally, the state-of-the-art UHF radio equipment with its new homing feature, which would enable pilots to rendezvous with tanker aircraft for refueling; and the Angle of Attack Indicator (AOL); and many other projects which have played their parts or are still playing their parts today.

A designated patrol plane commander, he served for four years as the commanding officer of a long range patrol and antisubmarine squadron flying Neptunes, known also by their designators P2V-5's and 7's, SP2E's, and finally as P-2's. He has four thousand hours of flight time in military aircraft, most of the time being in carrier aircraft types; and he has flown from seven aircraft carriers including the then-large CVB's *USS Midway*, and *USS Coral Sea*, the medium-size CV's *USS Wasp*, and *USS Tarawa*, the small or light CVL's *USS Wright* and *USS Saipan*, and the still smaller or escort carrier the CVE *USS Sicily*. His one hundred twenty-nine carrier landings have been in a variety of aircraft, both prop and jet types, and at all times of day and night. His experience includes the Navy's period of transition from props to jets and from straight-deck aircraft carriers to the angled-deck carriers designed for the jet age.

He flew in a regular Navy (USN) officer status for twelve years, during which time he received four commendations, most notably for a mercy flight he conducted, an investigation he pursued, and an electronics project he directed. Then he continued flying as a reserve Navy (USNR) officer for another twelve years. In other words, "he has been there, done that;" so he can write knowingly and personally about Naval Aviation. He writes humanely about Naval Aviation and its pilots, many, if not most of them in this book, he has known or talked with. He writes thoughtfully and with insight about the demands placed upon pilots flying powerful, sophisticated, state-of-the-art aircraft in challenging circumstances, often bringing out humor despite the difficult situations.

Author's Note

Knowing that there were many interesting and exciting stories to be told about Naval Aviation, it became a matter of how best to present them. Also, if the described incidents are to be of interest to the larger community of readers — that is, to others who are not pilots but who have an interest in flying and aviation — some brief explanations of flight essentials occasionally may prove helpful.

While the non-aviator may feel the excitement of Part I, in which Navy Pilots are descending or "letting down" through thousands and thousands of feet of solid clouds while experiencing fuel problems, overlooked would be the instrument flying skills and other requirements in these "letdowns." Crucial to the pilot is an understanding of essential elements of lift, stalls, etcetera if he is to remain in a safe flight attitude — not only during letdowns but also in other maneuvers. And absolutely essential are the pilot's discipline and self-control: the discipline to trust completely what his flight instruments are telling him and the self-control to exercise the patience required to persevere in very trying circumstances.

Any serious flying in today's state-of-the-art military or airline aircraft requires flying through bad weather or layers of clouds, to say nothing about turbulence, icing, or strong crosswinds on landing. Since most of these aircraft utilize turbo-jet engines, as opposed to turbo-prop engines whose requirements are somewhat similar, they must climb to and operate at high altitudes (35,000 to 40,000 feet or more) where the engines operate more efficiently and consume less fuel. To get to these high altitudes, more often than not, they must climb through layers or even solid clouds. How many days are cloud-free to enable the pilot to reach altitude instrument-free?

And if conditions are cloud-free at the takeoff point, what about the area enroute or at the destination, which could be hundreds of miles, or even thousands of miles distant? Add to the instrument flying required, the round-the-clock nature of military flying, which can be conducted in hostile environments or in pursuit of challenging missions.

And finally, the reader will note a wide range of incidents and challenges presented by the various types of aircraft used over a number of years, indicating the variety of occurrences in "routine training missions." Some will seem extremely demanding, while others are much less so; after all, these experiences are the nature of military flying and make up its sum. But even in those incidents which are less dangerous, most involve a decision made by either the pilot or someone else that affects the outcome of the flight. The pilot dealing with one problem, though serious, but who has lots of fuel and time and good weather, has to be compared with the pilot having multiple problems exacerbated by bad weather and a low fuel state. It is felt that presenting all levels of challenges will give the reader an appreciation of the diverse nature of what pilots deal with. After all, the number of extremely challenging, life-ending incidents is relatively rare. To focus only on those, to the exclusion of others, would result in a skewed presentation of Naval Aviation. Regardless of the severity of the situation requiring decisions, it is hoped that all narratives will prove interesting.

It is to be expected that Naval Aviators reading this book will react in some way, if only to wonder how others could find themselves in some of the incidents. But any experienced Naval Aviator will, no doubt, have experienced something similar or will have known of someone who had. The author welcomes, via the internet, reactions to these events or reports of similar incidents: <GandELeute@SDLifestyle.com> or <ana@durancreative.com>.

Included in the center section are photographs of some of the aircraft in which these incidents occurred.

The Hangar Doors Are Open

Whenever Naval Aviators, past, present, or both get together, it does not take too long before the topic gets around to describing and discussing incidents and accidents that each aviator has known through his experience, or from hearing about them from others, or from reading about them in the *Naval Aviation News* or *Approach*, both of which are largely flight safety magazines. No doubt, Air Force aviators have similar sessions and publications; and, most likely, airline pilots have them as well. The stories and descriptions are endless and would continue seemingly forever; for when you *Open the Hangar Doors*, they do not close of themselves. Some external pressure or necessity is required to close them — only to have them creep open again at the next gathering of a few Naval Aviators.

Naval Aviation has many interesting and exciting tales to tell. At one of the sessions mentioned above, the following incidents, for example, might be discussed: The pilot who landed with minus 250 pounds of fuel (minus 42 gallons); or the pilot who took off with his wings folded; or the pilot who had to land his transport aircraft with the wheels up because of mechanical problems — and didn't scratch the airplane; or the pilot who bailed out of his airplane only to have his parachute catch on the tail surfaces and be carried all the way to the ground — and amazingly to survival; or the aforementioned pilot who landed his big aircraft on a very icy runway in the Aleutian Islands and who, when the strong crosswind started turning him around so that, while his aircraft was moving in the direction of the runway, its nose was pointing more and more into the direction of the wind from the right that it seemed logical for him to add a little left engine power until he was pointed in the direction from which he came while moving backwards down the runway, and then add engine power to both engines to bring the plane to a stop, still pointing, of course, in the direction from which he came. Incidents like these come up often in aviator jam sessions; but there are many more, some of which will be presented in the following pages. Lest someone draw the conclusion that the incidents just described represent blunders or serious pilot error, let it be said that most represent reactions to conditions forced upon them, their responses reflecting credit on them by their reacting to difficult situations with skill, fast thinking, and courage.

The nature of serious flying in general and military flying in particular is such that, besides the flying skills required, the pilot is dealing with a host of variables and conditions over which he frequently has very little control but which he must face and react to. To start with, the military aircraft is an exceptionally complex machine with a strong airframe and powerful, amazingly reliable engines; but it has complicated systems which the pilot must know about and be able to manage: engine, fuel, oil, de-icing, navigation, hydraulics, electrical, electronics, weapons, and so on. Not only must the pilot know these systems thoroughly, he must also know emergency procedures for these systems and alternatives for managing them when problems crop up. Then, there is the military mission, which might require tree-top level tactics, abrupt high-speed maneuvers, and the operational imperative to get the job done — sometimes pressured all the more by a demanding, capable commanding officer determined that the mission be accomplished, at the same time thinking about advancing his own career. Overlying everything is the weather. While it is one thing to handle all of the complexities in daylight and good weather, it is quite something else at night, when the weather is bad, and there is the important requirement of military necessity. For such operations, an entire additional set of skills is required; and while the military equipment today is exceptionally good and exceptionally reliable, it is the pilot who must master the skills and have the drive and courage to make it all work in the manner in which it is intended.

In flying for the military, and particularly with pilots flying single-piloted, single engine operational aircraft, unexpected things occur. Sometimes they result in serious injury or death; other times, miraculously, no one is hurt or only slightly injured. With the wide range of incidents that can occur, it is not easy to classify them into groups. Only a few broad classifications seem relevant. Once again in these incidents, the reader, even without flying experience, may get a realistic feel for what the pilot is confronted with if he tries to put himself or herself in the pilot's seat.

With Naval Aviation, there is, in addition, the unique requirement for pilots to operate from aircraft carriers, giving Naval Aviation its own special character with its own set of required skills. The aircraft carrier has arresting cables for the aircraft's tail hook to catch on touchdown; it has barricades (or barriers) to arrest aircraft with hook problems, or bad approaches, or emergencies; it has powerful catapults to launch aircraft into the air in about two hundred feet of deck space; it has elevators to get aircraft up or down, to or from, the hangar deck and the flight deck; and it has landing signal officers (LSO's) and optical landing systems to direct the pilot's approach path to a landing on the flight deck. Besides the aircraft carrier's equipment, Naval aircraft have, as mentioned, tail hooks to catch one of the arresting cables and they have fittings or attachments for cables to connect them to the equipment which catapults them into the air with flying speed; they have wings that fold to conserve space in the limited environs of the aircraft carrier; and they have airframes especially designed for the ruggedness and rigors of carrier operations. Then, there are other requirements and factors peculiar to carrier operations: Landing intervals are important; "wave-offs" may be required for a fouled deck, an accident, some equipment difficulty, rough weather which causes excessive rolling or pitching of the carrier, or a bad landing approach by the pilot. Any of these wave-offs will necessitate more time, slowing down the efficiency of the process and making the carrier and its air group more vulnerable should an attack be imminent, not to mention that they will require the jet aircraft to use up precious fuel. Another requirement is that the pilot be able to find the carrier in the vast, endless expanse of ocean, a task made easier with today's advanced electronics, but made much more difficult if the pilot is experiencing electrical problems, navigational equipment difficulties, or if the carrier is operating in a "radio-silent" condition to avoid detection by a hostile force. Operating from an aircraft carrier requires flying at low altitudes (sometimes as low as 200 feet) and slow airspeeds in the landing pattern, a condition exacerbated by the need to make "approach turns" to the carrier during which more lift is required than level flight and which may require power adjustments to avoid stalling into the ocean. A further additional consideration for all Navy carrier pilots, but particularly for jet pilots, is the high fuel consumption at the low altitudes around the aircraft carrier. For the Naval Aviator, the need to make a good landing approach so that he can land on the first or second approach is crucial to save fuel — particularly in bad weather or when there are no alternate landing fields within reach of the aircraft carrier.

The complicating factor of bad weather presents special problems for the Naval Aviator and for aircraft carrier operations. Besides the unpredicted worsening of weather conditions, lingering conditions from a storm can present real challenges. Depending on the wind or on wave swells caused by a hurricane or a strong storm which has just passed through, the carrier can roll from side to side, it can pitch fore and aft causing the bow and stern to alternately rise and fall, and sometimes it can both pitch and roll at the same time. The pilot who approaches the stern of the carrier in good shape to receive a "cut signal" to land may find the stern sinking away from him, requiring him to make undesired last-second adjustments or else to wave-off and use extra fuel for another approach at a time when use of extra fuel may be critical. Nor is a rising deck caused by pitching any better, since an upward-moving deck and a descending aircraft may result in an increased rate of descent which could cause structural damage to the aircraft. As challenging as all of these conditions are, try to think about what could happen at an inopportune moment when an engine malfunctions or even "flames out," or an important system fails, or there's a fire in the aircraft, or the pilot is injured or wounded and is trying to bring an aircraft aboard that has some structural damage, all of which may be further complicated by bad weather. While only a few of these emergencies require drastic, immediate action such as "bailing out," all place additional stress and demands on the pilot and require that he think clearly and remain somewhat calm, despite the problems, to apply his skills and knowledge to best advantage. While some of these emergencies bring tragic consequences, the professional skills of the Naval Aviator most often prevail.

It is out of the unique and complicated nature of Naval Aviation and the skills — and sometimes the lapses — of the Naval Aviator that the incidents in this book are described. The public who have seen and heard reports of military pilots in operational flights during Desert Storm, Kosovo, or more recently in attacks on Afghanistan and Iraqi targets are aware that many flights have been at night and often in bad weather, yet they are executed with near surgical precision. Today, partly as a result of scientific developments, reliability of equipment, improved training, and increased demands placed on pilots, all-weather and round-the-clock operations are possible and even necessary. The news networks have shown pictures of jet aircraft being catapulted from the aircraft carrier into the black night in locations far from friendly airfields, requiring much of the pilot. He is launched into the blackness knowing that he has a mission to accomplish and that he must return safely back aboard the carrier — there is no other option. No one having this experience can honestly minimize it because, aside from flying skills, it requires courage. Today, the reliability of equipment and navigational devices is a great help; but things can and do happen, which, however you put it, can cause real concern and disastrous results

While to some of the public, it could appear that the difficulty and demands are overstated, they are not! To satisfy the demands of operational flying, pilots must have good instrument flying skills and must know about many weather conditions. Combined with this is another essential: to be thoroughly grounded in basic aerodynamics, especially the effects of air moving across the wings and other flight surfaces and the relationship of lift versus drag. Each is part of the same seamless cloth, weakness in one area affecting the whole. While all of these areas can be described in elaborate and complicated detail through the volumes that have been written about them, descriptions of the basics of these areas can be presented in rather simple, uncomplicated form. It would serve little purpose to go into these areas before any incidents have been related. Rather, it is sufficient to say that military flying, especially with today's sophisticated aircraft, requires the thorough integration of flying skills, instrument flying, and aerodynamic knowledge every day and for every flight undertaken by military pilots. For Naval Aviators, the effects of turns at slow speeds, as in an approach to the aircraft carrier, and an appreciation of the effects of torque when the power of a huge reciprocating engine is suddenly applied, and the slow response of jet engines to develop full power when the throttles are thrust full-forward are additional areas to be kept in mind when operating from aircraft carriers.

Part One: Fuel Starvation

Fuel Consumption and Weather — Twin Partners of Concern

Since most of the Navy aircraft today are turbojet-powered and the twin problems of fuel management and flying on instruments ever accompany jet aircraft, it seems a good place to start. The high rates of fuel consumption in jet military aircraft and the imperious nature of situations in which low fuel states occur place immediate and urgent demands on the pilot.

The predominance of jet-engined aircraft with their characteristics of high fuel consumption at low altitudes and much lower rates of consumption at high altitudes indicate that flight operations will be conducted mostly at high altitudes. Other than low altitudes around the aircraft carrier, tactical air strikes against land or ship targets, and fighter "dog fights" against enemy aircraft which descend to lower altitudes, there are not too many times when it is likely that jet military types will be operating at low altitudes or spending much time at low altitudes. Because there is the necessity to operate at high altitudes, another requirement is the need to climb through clouds or cloud layers to get "on top" of the clouds where most of the flight operations take place; indeed, some times the clouds extend solidly from just a few feet above the ground to thirty-five or forty thousand feet — and even above. Just the transition from the ground to the "on top" level requires that pilots have excellent skills in flying their aircraft completely with reference to the instruments in the cockpit since there may be no outside visual references. Even more skills are required for the instrument letdown and landing phase of the flight. While all of this is being described as a natural process and it will be taken for granted in the incidents in this part, these skills are the result of learning the techniques and practicing over and over again the processes involved, so that the pilot knows both that the system works and that he has the necessary confidence in himself.

But beyond the instrument flying aspect of jet military flying, there are other things involved. The pilot must know his aircraft's systems, not the least of which is the fuel system and its fuel consumption under various power settings and altitudes and how to dump fuel quickly in an emergency. Related to fuel consumption, is engine operation and the specific aircraft's peculiarities: How far back the throttle can be set during prolonged letdown? Is there, during the throttled-back state of the engine, a rough running or "rumble range" of the engine to be avoided? When and under what conditions should engine operation be taken out of the normal "automatic fuel metering" operation and placed in the emergency mode without flooding out the engine in what is known as a "rich out?" How long does it take for the engine to accelerate from a very low power setting to full power through the fuel metering device and how rapidly can the power be advanced when in the emergency mode? What are the procedures for restarting an engine that has been shut down purposely or has "flamed out?" And, if necessary, what are the procedures for utilizing the emergency restart cartridges, if available, to provide ignition to a flamed-out engine?

Although the reader need not know all of the technical aspects of these flights, he should be able to appreciate the challenging circumstances military pilots face, for the tension inherent in resolving the immediate problem in the following episodes in this chapter will be obvious — even if all the systems work properly and the pilot makes no mistakes.

Without going into too much detail, detail that could detract from the description of these episodes, knowledge of other systems is likewise necessary: familiarity with the hydraulic system and its use relative to the airframe is required, for example, because there are restrictions on when speed brakes and wing flaps can be employed, the extension of wheels and speed brakes having only limited restrictions at high speeds, while the use of wing flaps (or landing flaps) have severe restrictions. And what about a malfunctioning air conditioning system; in some emergencies, a troubling mist resembling smoke can come from the system, which, even if immediately recognized by the pilot as mist, is not the end of the problem — if it is necessary to turn off the air conditioning system, some times the Plexiglas canopy can ice over, completely eliminating or severely limiting visibility. There are other considerations as well, not the least of these being knowledge of emergency procedures involving bailing out. Although marvellous pilot ejection systems have been developed over the years, the pilot still has to know precise procedures if he is to avoid injury and be ejected safely from the aircraft. Much has been done recently to make ejection as foolproof as possible, but the pilot still has to make the decision to bail out — and in a timely fashion, since often there is little room for indecision and delay.

An attempt to group experiences has been made, although in fact there are similarities in almost all of the incidents presented in this book. However, even if the grouping is less than perfect, a starting point has to be made. For aircraft, running out of fuel is considerably more of a problem than in other fuel-burning vehicles, mainly because the pilot has to bring his aircraft down whether a runway is handy or not.

In the takeoff phase of a flight, fuel consumption in jet aircraft is unbelievable, the rate not greatly appreciated by people not involved in jet aviation. For example, one small 21,000 pound single-engined jet fighter aircraft like one of the Navy's relatively early jets, could, during takeoff, burn fuel at the rate of more than one thousand gallons per hour. That means that that same small jet aircraft could burn more fuel during forty-five minutes than the average 10,000 mile per year automobile driver burns in an entire year — assuming, of course, that the jet fighter pilot keeps his engine burning at takeoff power and keeps his aircraft at a low altitude where jet engines are inefficient. For larger aircraft and multi-engined aircraft doing the same thing, the rate of fuel use is staggering. Of course, keeping takeoff power and remaining at a low altitude is unlikely; but the point is that the consumption of fuel at takeoff, during climb, and while maneuvering tactically in dogfights is very high. It is a rate of consumption which has found its analogy in the rate at which a toilet flushes — in other words, the fuel really flushes out in such operations. And the analogy is a good one, because the toilet that flushes a couple of gallons of water in ten seconds would easily flush one thousand gallons or more per hour, a rate very close to our small jet fighter at takeoff. A twin-engined jet fighter like the Navy's F-14 Tomcat, F-18 Hornet, or any jet using an afterburner would boost the rate much higher. But even in times when the Navy's aircraft were powered by piston-fired reciprocating engines, there were fuel concerns, as an included event will emphasize.

Jet engines are designed for high-altitude use where they are much more efficient. The same small jet that burns 6,000 pounds (1,000 gallons) per hour at takeoff using one hundred percent of RPM can burn less than 1,500 pounds (250 gallons) per hour at high altitude using eighty-five percent of RPM. So, it is the takeoff, the climb to altitude at ninety-eight percent of RPM, and combat and tactical maneuvers that really burn fuel. And, of course, operating at low altitudes where fuel consumption is high is a real concern; for example, a pilot would not make a letdown through thirty-five thousand feet of clouds when the weather at the surface is below safe minimums for a landing. Rather, he would proceed to an alternate field and start his letdown there if the surface conditions were above unsafe levels. If the pilot let down where the weather was below minimums and then had to climb back to an efficient altitude, he might exhaust his fuel before he reached his alternate airfield. Another factor in all of this is that, ideally, a pilot should always land a jet with a certain reserve of fuel — about fifteen percent, or in the case of the small jet example, about one thousand pounds (roughly 166 gallons). That reserve is to insure that enough fuel would be available for some aircraft or runway problem that required the pilot to circle the field for a short period of time before he could land. But even for these types of contingencies, 166 gallons is precious little at that low altitude: a last-minute wave-off and a circle of the field once might consume a third of that reserve.

The impressive power and design of jet engines, not to mention aircraft design, are the reason for the very high aircraft speeds possible. The power of jet engines is measured in pounds of thrust. Although people tend to equate a pound of thrust with one horsepower, they really are equal only under certain conditions. Stated roughly, when an aircraft is traveling at three hundred twenty-six knots (373 miles per hour), each pound of thrust is equal to one horsepower, with varying relationships at other speeds. Consider the tremendous horsepower of the Jumbo aircraft types like the Boeing 747 or the Air Force's C-5 which have jet engines rated at about 45,000 pounds of thrust — quite a bit of horsepower and quite a bit of fuel consumption.

In reading about these incidents, the reader should probably think as the pilot does relative to the performance of the aircraft: of all the things that could possibly go wrong, the very high reliability of the aircraft and its systems insures that things are not likely to go wrong; but if they do, his training and knowledge should suffice.

The Flushing Toilet — Landing With Minus 250 Pounds of Fuel

Squadron Skipper Butcher and LT Givens' takeoff and climb to altitude were normal, climbing to the best possible combination of altitude and wind direction so that they could get the best range or distance from their fuel load. Though this was the usual procedure, this time it was more important since there was the possibility that they could make the trip from the airfield at Leeward Point, the larger of the two fields at the Naval Air Station Guantanamo Bay, Cuba, to the Naval Air Station, Atlantic City, without refueling. Leeward Point had been carved out of land across Guantanamo Bay west from the old, smaller NAS Guantanamo, where new runways could be built large enough to handle the increasing number of the Navy's jet aircraft. The older, more conveniently located field was still usable for certain types, but there was no practical way in which the runway lengths could have been lengthened. At Leeward Point, they could carry a full load of fuel to go as far as possible without refueling. Taking off one after the other and joining up in loose formation en route, they soon comfortably established themselves at altitude and on course in their F2H-2 Banshees; they then soon throttled back to a fuel-efficient cruising RPM and then shut one engine down to further conserve fuel. Shutting an engine down on this type of flight was not an uncommon practice. In fact, it served as good practice for restarting an engine in flight. They were prepared to refuel along the way if necessary, particularly since there were so many Naval Air Stations along their route. There was the Marine Air Station at Cherry Point, North Carolina, followed soon after by the fields in the Naval Air Station Norfolk complex at Oceana and Norfolk, Virginia. Oceana or Norfolk Naval Air Station would be the best places to refuel, if necessary. They had decided that, on nearing the Norfolk area, they would determine whether prudence would suggest refueling there. What went into the decision to press on without a stop was never publicized to the rest of the squadron, but they assessed their situation in the Norfolk area and decided to proceed, without stopping, directly to NAS Atlantic City — after all, being airborne and at altitude over Norfolk meant that the flight to NAS Atlantic City would be just a bit under twenty minutes duration. Besides, the weather was clear and they could still refuel at the Naval Air Station Chincoteague, Virginia, if they thought necessary or if an emergency arose.

Not quite ten minutes later and over Chincoteague, they could see the peninsula of Cape May clearly as Givens reported a fuel state of just 700 pounds (about 116 gallons). At this point, and figuring that it was as easy to letdown en route the remaining distance and just about as easy on fuel consumption as if they were to letdown to NAS Chincoteague, they decided to proceed on course to NAS Atlantic City. It should be remembered that both were excellent, experienced pilots making a judgment that their fuel consumption during their throttled-back engine settings would enable them to reach NAS Atlantic City safely — despite the requirement that they restart their shut down engine for landing and that they really should have been on the ground with about one thousand pounds (about 166 gallons) of fuel remaining.

Since Harry, as the wingman, had to make more throttle changes to maintain an appropriate wing position, those changes, however modest, required him to use more fuel than the skipper whose throttle setting remained constant. The result, of course, was that the Skipper's fuel state was somewhat better than Givens. So, with Cape May in sight and NAS Atlantic City just about thirty-five nautical miles up the peninsula, and the rest of the trip "down hill" so to speak with the throttles way back, they chose to go on. But at Chincoteague and using about fifteen hundred pounds per hour at altitude, Givens figured he might have less than one half hour's fuel left — provided he was not required to use more power at the lower altitudes they would soon approach. What happened next provides one of the best examples of Murphy's Law, or what can go wrong will go wrong — and at the worst possible moment! In other words, the rule for landing fuel minimums is there to allow for Murphy's Law. Good pilots, and these were good pilots, know why the requirement to land with a good reserve of fuel exists: to provide a margin of safety for unexpected contingencies like a temporarily closed field due to an accident, sudden changes in weather, etcetera, which may require them to proceed to another field.

   Already concerned about his fuel state and letting down rapidly, Givens' thoughts about his status caused a lapse in his concentration, allowing his position on the Skipper's wing to widen considerably. Ordinarily, this would not have become a problem so long as he could keep sight of him. But trouble began part way up the peninsula. A low overcast of clouds was encountered and somehow, in an instant, Givens lost sight of the Skipper's aircraft. The cloud cover seemed to come on them rather suddenly, and the Skipper went under the overcast a second or two before he did. In that second or two Harry became confused and thought that, if he looked around, he would surely see the other aircraft. Since Harry Givens had been flying wing on the Skipper, he had been relying on the Skipper's navigation. Now separated from the Skipper and extremely low on fuel, Givens had to shift to his own navigation — something which would require a little more time and use up some of the very precious fuel remaining. Within a moment or two the fuel counter indicator was reading "zero," and the field was not in sight. Now, Givens had to shift his thoughts to his imminent "engine flame-out" and the necessity of bailing-out, using the seat ejection system. There is an expression in Naval Aviation that, when a pilot is dealing with one emergency, it is not uncommon for a second, and perhaps even a third urgency to occur; and he is hemmed in or so overwhelmed by events closing in on him, that is "like a mousetrap starting to spring closed," and dictating the only option he has. This is how fast things happen, and now was such a time.

   Certain that he could not reach the air station even if it came into view, Harry called the Skipper, who was by this time circling the air station, to tell him that he had to bailout. The Skipper, with genuine concern in his voice, wished him well and advised him to be careful. Harry was now ready to pull the pre-ejection lever, an action that would jettison the Plexiglas cockpit canopy and move leg guards into position so that his legs would be close together and not be severely injured during his powerful ejection from the aircraft. Harry had already envisioned the next step, which was to reach up over his head with both hands, grab both handles of the face curtain and, keeping his elbows close to his body, pull the handles down hard in front of his face and chest till his hands were at belt level, at which time the big shotgun-like cartridge would fire to hurl him, his seat, and his parachute thirty or more feet safely above the aircraft. As Givens reached for the pre-ejection lever, he surprisingly saw the air station out of the corner of his eye and instinctively headed toward it. His plane was still flying, his engines still turning despite the fuel counter indicator's having been on empty for a few minutes. He resolved to keep going, but to be prepared to bailout — after all, it would take just an instant to pull the pre-ejection lever, pull the face curtain down and be in the clear. And in a worst possible situation, he could be ejected right through the Plexiglas cockpit canopy by eliminating the pre-ejection step and just pulling the face curtain down.

Givens received immediate clearance to land with a straight-in or direct approach from where he was, not knowing how he could still be in powered flight when he had passed the zero fuel level at what seemed like ages ago. But despite the tension and the confusion of so many thoughts during the last few minutes, Harry kept his cool and maintained his concentration as he crossed the airfield boundary, then the runway threshold, and touched down with what he thought was about <u>minus two hundred fifty pounds of fuel</u> (or minus about 42 gallons). The Skipper was exuberant at the outcome and much relieved, his relief exceeded only by that of LT Harry Givens. No doubt, there is an innate reluctance for a pilot to bailout while his aircraft is still flying and the engine is still running; it may be that this reluctance was a factor in Givens perseverance.

Everything that went through Givens' mind as he dealt with these problems is not likely to be known. Perhaps even he would have a difficult time in recalling every detail. What is known, for sure, is that he was able to function despite terribly distracting concerns and that he focused his intellect and skills successfully on the matter at hand to bring about the proper, if amazing, resolution of his difficulties. As to the matter of landing with minus two hundred fifty pounds of fuel, some might feel that prayer and a beneficent God played a hand. That may be so, but others naturally claim that it is impossible to fly with no fuel and look for more down-to-earth explanations. No doubt, Harry did pray! And perhaps the Good Lord helped him to function in a manner that enabled him to deal rationally and somehow free of panic with the very trying situation he was in. But Harry Givens, like many others, sought a logical reason to explain the impossibility of using two hundred fifty more pounds of fuel than he had.

Several possibilities could have accounted for the "apparent" use of more fuel than he had aboard, all of which center on the fuel gauge — or more properly the "fuel counter." Unlike a fuel gauge, which usually depends on probes or floats in the fuel tanks, a fuel counter measures pounds of fuel that flow through the counter (or a remote meter) en route to the engine fuel flow indicators in the cockpit. And unlike a fuel gauge, the fuel counter has to be set like the hand of a clock by the pilot himself before takeoff. One can easily imagine some, but not all, of the conditions which might arise, not the least of which would be forgetting to set the fuel counter at all. Although the pilot's takeoff check list would usually rule out the pilot's forgetting completely, it would not preclude his making an inaccurate setting. If the fuel tanks have been properly filled, then the pilot would set the counter accordingly. If the tanks had not been "topped off", an unusual situation, and the tanks were less than full, and the pilot, thinking they were full, set the counter for "full," the pilot would think he had more fuel than he actually had. But, if that had been Harry's situation, he would have run out of fuel while still thinking he had some fuel. More likely in the event of incorrectly setting the fuel counter is the possibility that the fuel counter was set at a lower than accurate position and his fuel flow counter, in counting the flow, reached zero while there was still fuel available.

Whatever caused Harry's problem — and it seems most likely that the fuel counter was unintentionally set incorrectly — a possible contributing factor has to be considered.

At the time of this incident, Navy jet pilots had to contend with two types of fuel: one was a kerosene-type known as JP-3, later superceded by more volatile JP's such as JP-4, JP-5, etcetera; the other was the high-octane gasoline for high compression piston-engined aircraft (octane rating 115/145). Jet plane fuel was available at shore stations; but aboard ship, only high-octane gasoline was available because there were still some piston types of aircraft which used only such fuel; and aircraft carriers at this time did not have separate storage tanks for the two types of fuel. While jets could use both types of fuel, piston types definitely could only use the high-octane fuel. And while the ship storage problem may not seem to present any real difficulty, in reality it presented the jet pilot with one more thing to keep in mind: that the kerosene fuel is more dense, hence heavier, and since the fuel counter "measures weight rather than volume," this must be considered when setting the fuel counter. So it is quite possible that Givens, having a fuel load of kerosene, thought he had gasoline and set the fuel counter for the lighter gasoline rather than the heavier kerosene fuel he had received from the Leeward Point airfield. If the fuel had been gasoline, Harry really would have been out of fuel and would have had to eject — but the heavier kerosene likely provided his unexpected reserve — but opinions varied as is to be expected.

On most flights at that time the pilot was required to have from one thousand to twelve hundred pounds of fuel on landing, roughly 166 to 200 gallons. If the pilot lands with that reserve, the fuel counter setting according to whether he has JP fuel or gasoline is not critical — though still important, because the duration of jet fighter flights without air-to-air refueling is brief at best. Many times when weather and other considerations are good, the recommended reserve of fuel for landing is shaved a bit by pilots. Even such shaving is risky, and Naval Air Training and Operating Procedures (NATOPS) are established for good reasons, as this flight would indicate. Be that as it may, however, LT Harry Givens could have reached the state he was in from entirely different, legitimate reasons; and his reaction and actions could hardly have been subject to criticism. In Givens case, no doubt, it had been the combination of stretching things a bit and an inaccurate fuel counter setting that created the problem, regardless of whether the type of fuel played a part. Nevertheless, those last harrowing minutes and seconds of LT Harry Givens' flight were real, and he acted with courage and in a professional manner — even if the question remains about how experienced Naval Aviators place themselves in such positions just to save the hour or less that a letdown and refueling would require. If everything goes smoothly, it's fine, but the potential for real trouble gets built-in to the process if a safety margin is not provided for.

A full accounting of the details of this incident was never presented or discussed at the daily safety briefing the next morning. For obvious reasons, neither the Skipper nor Givens wanted to emphasize the misjudgments; and since no one was hurt, they felt it best not to advertise. There would be another day, another day of flying, a day when they would be a little wiser. Like many experiences, people find a need to talk about them. It was that way with LT Harry Givens who felt a need, certainly understandable, and some relief in sharing his incident with us, squadron mates who would not be judgmental, realizing, of course, that next time it could be any one of us. LT Givens' description of his incident was calm and matter-of-fact, void of any sense of excitement or emotion. After all, tomorrow would be another day with another flight or two, and he would be right in there.

Should the reader think that fuel-related occurrences like this are isolated incidents? They are not, but neither are they common experiences. Somewhat similar incidents, yet quite different in origin, follow.

The Flushing Toilet — Plus Bad Weather

If the preceding incident seemed unique, fuel problems in military jets, particularly fighter and attack types of aircraft, are not unusual. It is the nature of turbo jet aircraft and their rapid consumption of fuel that they occur with some frequency — even when the pilot feels sure he is using conservative procedures. Nor are fuel problems only a concern in military jet aircraft. The January 1990 crash of a Boeing 707, Avianca Airlines Flight 52 out of Colombia, South America, was the result of fuel starvation. Although the pilots did report a low fuel state, there was some question about whether they used the proper English terminology to report that fuel state. They were cleared to make an approach to Kennedy, but their approach was such that they had to abort the approach and attempt another. Ten minutes later, while waiting to be cleared for another approach, the 707, now out of fuel, crashed near the North Shore — with multiple casualties. Although the Avianca crash is too complicated and too detailed to go into here, one aspect deserves mention: being down to a very low fuel state when the pilots were granted clearance to make an approach to the runway, they did not get a second chance, because the go-around with their wheels and flaps down required a lot of power and, of course, a high rate of fuel consumption.

An interesting incident, which occurred when an eight plane flight of two four-plane divisions of F9F-6 Cougars was returning to base at an East Coast Naval Air Station one morning, provided a very tense situation. At takeoff time two hours earlier (about 0900), the weather had been poor; and as the aircraft were returning to base, the cloud ceiling was reported to be 400 feet and the visibility to be a relatively good three miles. Since the cloud cover extended to about thirty-five thousand feet, a so called "flat-top" type of descent through the extensive cloud cover was decided upon, at the bottom of which, and as each pilot flew over a radio beacon, his aircraft would, in turn, be picked up by radar for a precision approach to the runway through the remaining layers of clouds and be guided in to the runway. The radar was part of the base's Ground Controlled Approach system commonly called GCA. The "flat-top" approach was somewhat similar to the manner in which a division or two or more aircraft would fly by the starboard side of an aircraft carrier headed upwind and break off individually after brief intervals of five to ten seconds. They would make a one hundred eighty degree turn, heading downwind abeam the port side of the carrier, where, at a point even with the stern of the ship, each plane would start its turn in to the carrier for a landing. The big difference in these maneuvers is that, the break-off turn upwind of the carrier is level at about 500 feet, and only a slight descent is required downwind in the approach turn to the carrier deck which is roughly sixty-five feet above the water.

This "flat-top" approach required the use of an omnidirectional beacon, from which (outbound) and to which (inbound) the pilot would maintain specific bearings. In this type of group letdown, it was always preferred to be either the flight leader, or if not the leader, at least part of the first division. But this day George Brown found himself the last plane in the second division. Not really concerned about his position because he had gone through this type of letdown many times, the thought of complications occurring when eight aircraft were in such close order did cross his mind.

The eight-plane flight approached the beacon from the west at 35,000 feet; and crossing the beacon, it maintained level flight for a couple of minutes until the aircraft broke off individually with fifteen to twenty second intervals. At this altitude, strong westerly winds could be counted on, so there was no need to fly easterly for more than a couple of minutes past the beacon before the lead plane could start his turn. The turn would be a one hundred eighty degree descending left turn to the direction of the beacon, now to the west of the aircraft. The first plane was followed at short intervals by the seven others in carrier-like manner. The idea was to make the turn, letting down at 4,000 feet per minute and head to the beacon in single file, descending all the way to the same beacon they had crossed in an easterly direction at 35,000 feet. This time, they were to cross the beacon inbound at 2,000 feet, which they were to maintain till advised. Ideally, on reaching the beacon inbound, each plane would contact and be picked up by the GCA controllers operating the radar and voice communications. GCA had several frequencies so the controllers could work with a number of aircraft at the same time. As one plane would touch down on the runway, that frequency was available for another aircraft farther behind. It was a good system, enabling quite a few aircraft to be brought in during bad weather in a relatively short time — provided everything went as intended to make the system work. Naturally, despite the best laid plans, things do not always work out as planned or intended — again Murphy's Law.

The four aircraft in the first division were brought in "without incident" — that is, the system worked so far as it was supposed to work. If the second division came in as well, all eight aircraft would have been brought in to landings in something like twelve minutes. But unknown to Brown immediately was an incident which occurred when the first aircraft in the second division blew a tire on landing, and the runway on which GCA was set up was closed. Already in contact with GCA after crossing the low beacon, Brown first learned about it when he received a general announcement that the remaining three aircraft would be vectored to another runway with what was called a "Planned Precision Approach" (often referred to as a PPI or PPA) instead of the regular, more precise GCA approach. Naturally, the PPI was second best, one that was approximate rather than specific. Essentially, the PPI approach was still a radar approach, but one intended just to lead you over the field at a certain altitude, not to bring you in on the runway heading with specific letdown instructions all the way down to the runway. But a PPI represented the best that could be done under the circumstances. It did not seem likely that the GCA unit could be set up very quickly on another runway, and perhaps the original runway could be cleared in a couple of minutes anyway — not so! The news that they would be brought in using PPI's caused not the slightest concern in the cockpits of the three remaining aircraft. All the pilots had experienced them before and it was no big deal: all it meant was that, though less precise without the GCA azimuth headings and glide slope, each plane would be given a series of vectors in turn to bring it over the air station, under the clouds, and in the general vicinity of the runway to be used. The pilot would then use his discretion and, since he would be under the clouds, take over visual control of his approach and landing. However, a major concern when jet aircraft are nearing the flight's conclusion and are low on fuel anyway is that the fuel is rapidly disappearing at the now assigned low altitudes. There is no where the pilot can go — he hasn't enough fuel. He cannot be refueled. He is stuck with the system.

In theory, the plan and procedure seemed fairly good — not always so in practice, particularly when there were several aircraft to be brought in. The big problem was speed, time, fuel, and possibly radio communications problems. It took only about a minute or so for the aircraft that crossed the low beacon at 2,000 feet to reach the air station even at relatively slow flight; and precious time had been lost in switching over to the PPI procedure. Since the remaining three aircraft would be on different GCA (now PPI) frequencies after crossing the beacon, each pilot did not know how much progress was being made with the aircraft in front of him; he only knew that he was to maintain his current heading and that he would be contacted on the assigned frequency. George Brown, having earlier been cleared down to four hundred feet, getting low on fuel and now past and heading well away from the air station was wondering how soon GCA would call him. Solidly in the clouds, concern grew that he was getting farther and farther away from the air station, that there were some high radio towers in the vicinity, and that it was likely that, at this low altitude, radio communications could soon be difficult if not impossible. Now down to 800 pounds of fuel (133 gallons) and getting close to the point (600 pounds) where the low level fuel warning light would come on, he was starting to think of alternatives, uppermost being that he would climb to one thousand feet and head back in the direction of the beacon. It might take him over the field and, at least, it would improve radio reception. At worst, he could go to the beacon and take an inbound heading to the field, letting down till he was under the clouds. As he wondered whether that alternative would work, yet disciplined to wait a bit longer, a beautiful call, loud an clear, came over his radio from the GCA people. It was nice to know that he had not been forgotten and that he was still, at his altitude, in radio range. Although he did not think about it until later, it dawned on Brown that, if he were getting itchy and a bit worried, so too must have been the operators in the GCA hut.

Brown relaxed, thinking his problems were over. He knew now that they were concentrating on him, the last of the eight plane flight, and that he would soon be on the ground with some reserve of fuel. Brown was given an immediate heading change and was advised that he would be vectored over the field and he would have to make his own approach to the vicinity of the runway. GCA reported good visibility under the clouds and estimated that he would be beneath the clouds at 400 feet — not a bad deal, since all the pilots had known much worse at one time or another. Turning to the assigned heading of 110 degrees and adjusting it slightly as he maintained his 400 feet of altitude, he thought it was such a good feeling to know that you were a blip on some radar operator's scope and that he was working to help bring you in. At about the same time that George Brown realized that, at 400 feet, he was not under the clouds but in the clouds, a yellow flashing light on the instrument panel caught his attention, as flashing cockpit lights always get your attention, to advise him that he had reached the low fuel level of 600 pounds (100 gallons) — not exactly a real worry when you know that you are near the air station and GCA people are working you. Another call advised George that he was over the air station and should take over visually for his landing approach — but George could see nothing below. Just when Brown thought there were no further complications, he realized that there were. Not only was the ceiling lower than 400 feet, but there was a lower level of scattered to broken clouds at 250 feet The term "scattered clouds" refers to a level of clouds that covers less than half the sky and does not constitute a ceiling. The term "broken clouds" refers to clouds that cover half or more than half of the sky and do constitute a ceiling. In many cases, the distinction is not worth making; but in this case, the broken clouds at 400 feet very much limited what Brown could see below and he had no forward visibility in the clouds. Brown letdown a bit to 300 feet and, between the scattered clouds, could see the ground — but no air station! GCA continued the vectors with one followed by another and that followed by another; and it now seemed likely to Brown that the initial heading had brought him to the right of the air station and that he was circling the field but just out of sight of it. He was reluctant to go below 300 feet when he was not on a precision approach and did not yet have sight of the air station.

On the third or fourth close-in heading change, he was down to 250 feet and 400 pounds of fuel (67 gallons), when, through the broken clouds, he caught a glimpse of the edge of the air station and immediately dropped down to 200 feet where he was completely clear of the clouds and, for the first time, saw the entire field. As Brown broke clear and below the scattered clouds, he was advised that he was cleared for an immediate landing on runway twenty-seven, heading west. He would have preferred a closer runway so that he would not have to use more fuel to circle nearly half of the field. But he was not complaining, only thinking what an unusual picture he presented flying so unusually close to the ground for so long as he flew around to his assigned runway. The result was that he made an aircraft carrier approach and a very emotionally soothing landing on Runway 27. If Brown was at all affected by his experiences involving fuel, weather, and GCA problems this day, his approach and landing did not seem to show it. When LT George Brown turned off the runway, he looked at his fuel counter to observe that he had just a bit over 275 pounds of fuel (49 gallons) left — perhaps enough for a wave-off, a circle of the field, and another landing approach provided all of the fuel was usable. But George Brown was just as well pleased that such was not necessary — you can be sure!

Eight pilots had been involved, three of whose approaches could not be processed in the routine manner by GCA and had to be handled by the alternative, less precise PPI method. It would seem that some debriefing of the entire flight would have been helpful, if even for the pilots to share their approach experiences. That did not occur, because eight planes had been brought in, however safely, and the press of other duties took precedence. But if George Brown was disappointed or annoyed by the lack of attention to this flight, it certainly did not bother him. Like everyone else he was left to put it all together, learn from it, and get on to other things.

The Flushing Toilet — Plus Bad Weather and Radio Problems

Simply, just a routine night training mission. But simple routine training missions can mask some exciting experiences that hardly anyone gets to know about. Indeed, many people might not really care about them. That seemed to be the case with this one. But the eight plane flight of jet aircraft that took off from an East Coast Naval Air Station had at least one experience of note. The two four-plane divisions climbed single-file up through a very solid overcast that extended from two thousand feet up to about thirty thousand feet and joined up in the clear skies above the overcast to pursue the mission. The first division consisted of four F2H-3 Banshees, considerably upgraded versions of the F2H-2 Banshee series: and four F9F-6 Cougars. When the mission was completed, the flight broke up into four two-plane sections to make the best use of the time remaining — that is, the best use of the fuel remaining. At this point, the sections of Cougars had roughly 1400 pounds of fuel remaining from the roughly 6000 pounds of fuel at takeoff. Actually, the aircraft had just about the right amount of fuel left to begin an approved descent through the thick cloud layer and return to base. By leaving at this time, they would be relatively close but less than the recommended 1000 pounds of fuel when they were actually on the ground. It was certain that both pilots would be of similar thinking on the desired fuel state on landing and the time needed, hence fuel required, for return to base. But in this case at this time, Wilson seemed to be dawdling, making no attempt to head for the point from which a letdown through the thick overcast would be made. Now, for some unknown reason, radio contact between the two aircraft ceased to exist. LT Al Binns, who was flying wing on LCDR George Wilson, had not switched from the tactical frequency in use, nor should he have switched. So he knew that, whatever the problem,

Fuel Starvation 21

it was not his doing. He tried to call Wilson on the tactical frequency several times to no avail; but he did think that George Wilson would soon be back on "tactical" and would give him a call. Binns was thinking that Wilson had temporarily switched to another frequency to request letdown instructions. Since it was a dark night without a moon, LT Binns decided to concentrate on his close-in formation flying, watch for any hand signals from Wilson indicating a problem, and, if not, just wait for a call. Also, Binns felt that they would soon be making a letdown through the overcast in formation — not a common nor necessarily safe letdown procedure — so he stayed close.

What caused the loss of radio contact was not known. Since Binns had not changed his frequency or even touched his radio controls, and since he was receiving local beacons and navigational frequencies loud and clear, he assumed that, whatever the problem was, it was the section leader's problem and that he would soon resolve it. Since LCDR Wilson was one of the most experienced and capable pilots in the squadron, LT Binns' confidence in his leader did not seem to be misplaced. Was this well-placed or misplaced confidence? That is difficult to say. At any rate, Binns felt sure that Wilson had everything under control, and all he had to do was to fly on his wing. But what Al Binns did not know, in fact could not really know under the circumstances, was that LCDR Wilson was lost or that he, at least, did not know exactly where he was. Had there been radio contact between them and had Wilson asked Binns to take over for navigation and letdown through the clouds, there would have been no problem since Binns likewise was a skilled and experienced pilot.

Binns was just about to call Wilson on "guard frequency," an emergency frequency of 243.0 megacycles and a frequency that every pilot should be "guarding," that is, listening to in addition to any other frequency being used. But before he could call, he saw Wilson start to descend through the clouds. Since Binns had not heard any voice transmissions, he was a bit surprised at the start of the descent. So his attempt to call his section leader on "guard" had to be postponed; and as things followed one after the other, he would not again get the opportunity to call on "guard" frequency, all of his attention having to be focused on maintaining a close wing position. At this time, Binns fuel level was about 800 pounds. The descent to get below the anticipated ceiling of 2000 feet, letting down at the standard rate of descent at 4000 feet per minute, would require about seven or possibly eight minutes; and even at the much-reduced power setting during the descent, some fuel would be consumed. But the more pressing, immediate concern for Binns was to fly close enough to the leader during the descent to keep him in sight in the dense clouds and at night. Even though the leader's wing-tip navigation lights were on and the burning exhaust from the leader's tail pipe could be seen, this task was not easy. Visibility in these circumstances was very bad, even the flame coming out of the leader's tail pipe was an unexpectedly dull yellow glow, not the bright orange that you would expect; and the dark blue paint on the leader's aircraft made the shape of the aircraft seem to blend in with the dark of the clouds. To keep LCDR Wilson's aircraft in sight, Binns had to stay very close, a position that increased the risk of collision; yet each time that Wilson adjusted his throttle setting, as was necessary and expected, LT Binns was in danger of overriding his leader. Although the danger of a midair collision was slight, it was, nevertheless, a possibility. Should Binns move far enough up or abreast of Wilson, he could lose sight of the exhaust flame. Losing sight of the leader's exhaust flame would mean, of course, that he had lost sight of the leader's aircraft. Such an event would leave both aircraft letting down very close to one another without either one seeing the other. Some kind of turn, if only a zig and a zag by the wingman, would have been required to create some separation. But what if Binns and Wilson made the same turns? That was highly unlikely since the leader would turn to the left, and his wingman, flying on the leader's right side, would turn right.

The thought that he might lose sight of Wilson made Binns all the more determined to maintain his proper relative position and keep the leader in sight. It was, by far, the more preferred course of action whatever the effort required, whatever the concentration required, and whatever the throttle adjustments required. Now, four minutes into the descent, a big bright yellow light came on LT Binns' control panel, its brightness intensified by the darkness of the thick clouds they were in, and made Wilson's exhaust yellow even more dim. The bright yellow light was a fuel "low level" warning light which comes on at the 600 pounds level to warn Binns that his fuel was getting low in case Binns did not already know about it — fat chance! Now going below 600, and with several more minutes of descent left, it was a situation that was getting tight but, of course, not yet critical. LT Binns felt that he had no choice but to keep faith in his section leader as the best and most expedient solution to the problems that were building up.

Now descending through 6000 feet, LT Binns felt sure they would, in a minute or so, break through the overcast; and if the visibility remained as good as it was at takeoff two hours ago, they would complete the flight without any further problems. Sure enough, in less than a minute, both aircraft had broken beneath the clouds into the good visibility and bright lights of the Eastern Coastline. Although LCDR Wilson's letdown procedure lacked the crisp precision desired and expected, at breakthrough they were about three or four miles northeast of the air station, now clearly in sight. Instinctively, Al Binns switched to control tower frequency, surprisingly just in time to hear Wilson call for landing clearance. Knowing his wingman would have less fuel that he had, Wilson, using the control tower frequency, asked Binns for his fuel state. On hearing Binns' report, a fuel state now at 400 pounds, and having been given Runway 18 or 180 degrees for landing, Wilson told Binns to make a straight-in approach from where they were rather than the normal circling, more time and fuel-consuming approach. Relieved at the sight of the airport, the good visibility, the landing clearance, and the prospect of a straight-in approach, Binns made straight for the air base, now feeling relatively relieved that the situation was manageable. But he was realistic enough to know that problems could still crop up if he should make a poor landing approach or an unexpected runway obstruction required him to take a wave-off and go around for another approach. But Binns was sure he would not make a bad approach, not only because he could not afford to but also because he knew he would be very attentive to detail — and a possible runway obstruction seemed remote at this time of night. As he expected, Binns' approach and landing were right on the money, any tension or concern not in evidence.

While Binns was making his approach, Wilson watched Binns until Binns was in his final approach position; and, being satisfied that everything would be okay, Wilson made the normal circling approach to a landing. Once safely on the ground and having taxied off the active runway, each pilot checked his fuel state: Binns had 250 pounds, and Wilson had 400 pounds. Had all the fuel in Binns' tank been usable, he might possibly have been able to make a normal circling approach to the runway; however, had the runway been obstructed by a taxiing aircraft or a truck whose driver had become disoriented, it is doubtful that LT Binns could have taken a wave-off and had enough fuel for a go-around and another landing attempt, particularly since the low altitude and the throttle setting now created a situation that makes fuel disappear down that "flushing toilet."

Binns was fairly certain that Wilson's letdown was off-the-mark because their position when breaking under the clouds was too far from the air field. Binns wondered if the letdown had been a guess or estimate, even thinking that Wilson might have used a different beacon or that Wilson had just strayed too far off the precise course. Later, Binns felt that he never really received a satisfactory explanation of the events which Wilson really owed him, because as his wingman, he had experienced some unknowns, raising many questions. But there was no need to think about that now. Wilson, whatever the undisclosed problems he had experienced, had made some judgments, judgments which provided a satisfactory resolution. Both pilots knew that fuel concerns were always something to be dealt with; and both pilots, like jet pilots everywhere, know that these types of experiences are not exactly unknown. While it might be thought that Wilson and Binns' situation was tense — and it was — it was not a series of blunders but rather undetermined radio problems that created their incident. On the other hand, it was their skill, judgment, and calm perseverance which brought about the timely and successful resolution of their difficult situation.

Less than forty-five minutes later, both Wilson and Binns were at home with their wives and children. Wilson had three whisky and sodas to unwind; and Al Binns settled in to what remained of a quiet evening with the wife and the act of putting his small children to bed. Neither pilot placed any undue importance on the earlier events of the evening, thinking more about a little time to relax before getting a half decent night's sleep so that they could be ready for tomorrow's flight responsibilities. They were, nevertheless, very much aware that situations like their's this evening were close to the edge of a real life-threatening incident, a situation to learn from and not to be repeated — much less to be dwelt upon. Wilson and Binns were both scheduled for a couple of flights the next day, one being an early morning flight — both pilots knew instinctively that none of tomorrow's flights would be as "interesting" as this night's — after all, such occurrences are very rare; and tomorrow would be another day. Somewhat humorously, as he drove home from the base, it had occurred to Binns that what had just taken place on the last flight of the day was like what some people experience at the end of "a bad day at the office" — no better, no worse! He knew that there would certainly be good "office" days to come.

The Flushing Toilet — Plus Bad Weather and Both Engines Shutdown

While the reader might assume, not altogether unreasonably, that descriptions of the incidents in this chapter are arranged in order of the difficulty of the challenges faced, that is not necessarily the case since each incident has its own set of difficult challenges. So before leaving this area of close encounters with empty fuel tanks, it might be worthwhile to include briefly just one more incident of incredible achievement in an situation no writer of fiction would be likely to invent. The source of the information on this incident comes from the *Naval Aviation News*, a bimonthly magazine published for the Chief of Naval Operations by the Naval Historical Center, and is described by Grampaw Pettibone who has been mentioned earlier.

This incident is framed in a positive light, yet it ends leaving many questions. The reader is asked to judge the incident on what the pilot was able to do from the time the incident begins; that is, when the pilot was extremely low on fuel as he neared the vicinity of Memphis, Tennessee, in his F7U Cutlass at high altitude and in very bad weather. At about thirty thousand feet altitude and already having shut one engine down to conserve fuel, he received clearance to descend through the overcast for a landing. Knowing that, at a standard rate of descent of 4,000 feet per minute, his descent would take more than seven minutes and thinking he well might run out of fuel before breaking below the low clouds, he decided to shut down his remaining engine and descend without power for several minutes. So, at this point, he is in the clouds trying to concentrate on his instruments to make a safe descent and a good approach. He is perilously near being out of fuel; and, by choice, he has no engine power. He is hoping that, when he attempts to restart an engine, it will restart; and he is praying that, if it restarts, he'll be able to make the field and land safely before running out of fuel. At this point, many pilots would also be considering their bailout procedures — and, no doubt, this was also in the Cutlass' pilot's mind. But with a mind now marvelously concentrated, he carried on; and after descending for about six minutes and carefully checking his altimeter, he restarted an engine. Shortly thereafter, he broke below the clouds and made it safely to the field.

No pilot whom this author knows who is familiar with this incident has anything but admiration for the skills this Naval Aviator employed so admirably or had anything but praise for the concentration, for the cool manner of execution, and for the courage he exhibited in the circumstances. Is it any wonder that Gramdpaw Pettibone used some of his best exclamations in learning about this incident.

Although many questions arise in examining the circumstances leading up to and after this incident, no judgments or comments are necessary here, except to say that the pilot like most of the others in this chapter had to share his story.

Setting aside this incident of the F7U Cutlass, consider some of the sources of the problems the pilots encountered. In the first episode, the lead pilot pushed on to a questionable degree to avoid refueling; his decision caused a near bailout condition for his wingman, who hardly could question his skipper's decision. In the next incident, a blown tire resulted in a runway's being closed at a time when eight aircraft were being brought down in close time sequence during a period of low clouds. In the third, radio difficulties contributed to but did not cause the problem. Most serious was the lead pilot's not planning or not deciding to land with the proper reserve of fuel. But also in the third incident, there was what had to be considered an imprecise letdown through many thousands of feet of dense clouds, although it seems possible that the lead pilot may have just letdown in an area he thought was in close proximity to the air station. In all three cases, a pilot who had nothing to do with decisions which placed him in such a demanding place found himself in a heap of trouble from causes over which he had very little or no control.

It is fair to ask what it is that makes these pilots accept incidents like these with a fair degree of equanimity. Certainly they are not automatons, lacking intellect, fear, or reason. Nor are they stoic or fatalistic. On the contrary, they are realistic and have seen some of their fellow pilots lose their lives in aircraft accidents. They do not persist in thinking that deadly accidents can only occur to other people. They are young men, mostly, with wives, families, and the mundane concerns of schools, communities, and paying bills. It's just that they have chosen a career and employment so relatively few have chosen. But perhaps they are unusual in that they have learned to put a near-miss in the past and, as they say, "get on with it." And their pride which will not let them be less than what is required. Naval Aviation is voluntary; a pilot can give up his wings at any time he chooses. That some pilots have turned in their wings after particularly trying experiences is evidence of that. But, while many have nearly reached that point, even saying, "I quit," the actual number is very small. Usually it is other non-threatening reasons that cause that decision.

That many Naval Aviators flying carrier aircraft for some time have had similar fuel problems can be attested to by many others, including Captain Dale "Snort" Snodgrass USN, who has 4600 hours in the F-14 Tomcat and hundreds and hundreds of carrier landings in his twenty-six years of flying. As the Fighter Wing commander of the Atlantic Fleet, he was probably the most experienced Tomcat pilot the Navy has had. His description of his own very tense incident is related in "Tomcat Tales," a publication of the F-14 Association. His brief, blunt description makes the point impressively:

"One night, I was very low on fuel, the weather was terrible, and the (carrier) deck was moving 15 - 20 feet. I was waived off twice because of a fouled deck... something that had nothing to do with me. I boltered (touched down just past the wires) on my third try and went around for a fourth time. It was a real ugly pass... the deck was moving a lot and I was feeling more and more stress due to my rapidly dwindling fuel supply and no tanker (overhead for refueling). I got on the deck with less than five minutes fuel remaining. *The thing is, though... almost every carrier aviator has a story like this.* Ask any carrier aviator and they will tell you, life at the back end of a ship is one part thrill, one part chaos, and one part stress.

Only the strong survive. It's simply the toughest environment in existence."

Would five minutes of fuel have allowed another pass? What do you think?

Part Two: Where There's a Will

Where There's A Will — The Mission Is All

Not all incidents in Naval Aviation have the immediacy, urgency, or threat of injury or impending death as mentioned in Chapter One. Nevertheless, in the sum of "routine training flights," things happen which also require judgments followed by decisions which affect the flight's outcome. Nowhere is this more likely to occur than in trying to fulfill an operational mission. After all, the operational mission is really the reason or justification of military aviation.

The idea that "The Mission Is All Important" is one that can mean many things to many people, the range of differences being largely due to background and experience. One need not have a military background to appreciate its significance. For the ill-informed or less widely read person, it may have a limited meaning. And sometimes a person's antimilitary bias puts a strange, usually equally ill-informed spin on his views; to that person, it is a rash, ill-considered, do-it-at-any-cost opinion of the "I love any kind of war" military mentality. Nothing could be further from the truth. One Pulitzer prize-winning cartoonist always draws pictures of high ranking military officers as fat persons meeting furtively. To put a kind face on his shortcomings, he exhibits at least a two-sided ignorance: one about the military in general, and the other about the character and appearance of its officers; the senior officers are not secretly planning the next war somewhere — on the contrary, far from seeking the next war, military officers want to avoid wars; but they also take seriously their responsibility to maintain readiness for conflict should it occur — and fat senior officers are as scarce as hens' teeth. While he and others are entitled to have their own opinions, there is something to be said for a reporter or cartoonist' knowing at least a little bit about his subject and having some sense of responsibility to the profession he represents and the public he serves.

The mission is very important! It is the heart and reason for whatever is being attempted. It is the focus of the pilot's efforts. All planning leads up to its accomplishment. Safety, undue risk, and the likelihood of achieving mission fulfillment must be taken into account. On the other hand, there is always risk. But a timid, weak, or unenthusiastic effort is unthinkable. Wing commanders know their pilots capabilities and stress aggressive action as the means of getting the job done when you are in "harm's way." Former Defense Secretary William Perry stressed this idea after the terroristic bombing of U.S. military housing in Saudi Arabia when he said that, in spite of dangers, U.S. forces were there to carry out a mission and that some risk is inherent. And the Battle of Midway (June 4, 1942), the greatest Naval victory in the Pacific, might never have occurred as it did if the sound thinking, planning, and aggressive action of Admiral Raymond L. Spruance USN had been tempered with less confidence in his people and equipment and more caution and reticence. In the Battle of Midway, U.S. military personnel were indeed killed, some Naval Aviation squadrons being completely destroyed or severely weakened. But surely the destruction of four major Japanese aircraft carriers and the elimination of Japan as a real naval force in the Pacific at that time saved thousands of lives.

Mission awareness is a part of all military flights. Without dwelling on the concept of "mission," and indeed it is not a much discussed topic, officers and crew learn very early on that the "mission" assumes the central and overriding consideration. "Mission" strengthens and enhances the ideas of cooperation and leadership: both learning to follow responsible leaders and developing stronger leadership in junior officers. The importance of getting the job done well is a part of every flight whether it is some practical, utilitarian flight such as flying to join an aircraft carrier already at sea or delivering a plane to some other airfield. The same approach and degree of preparation become a part of every flight.

Parts of any mission, in addition to carrying ordnance and knowing how to operate weapons systems, are those ongoing requirements of navigation, instrument flight, fuel management, contending with radio and mechanical aberrations, and, of course, dealing with weariness. Problems with any of these bring an awareness that danger can be a part of any flight. But to get the job done, these potential problems are relegated to the bottom of more pressing concerns to be faced only if and when they occur. Indeed, this is rational and proper because serious problems are rare; and to place undue emphasis on unlikely events would be seriously limiting. But these problems and concerns can occur, however infrequently, on any flight whether in jet or prop types, though fuel problems are perceived as more pressing in jet types — a perception perhaps strengthened by the incidents in Chapter One. But propeller types have fuel problems too, as the following episodes reveal — not only fuel problems but also many other problems.

The Simple Mission? No Way!

Although a very long assignment, the mission seemed simple enough: As part of a very extensive test to determine whether attacking aircraft could avoid radar detection by flying to designated targets at very low altitudes. The thinking was that, although detection would be possible at some point, the aircraft would be so close to the assigned targets that they could carry out their assigned missions. Higher altitudes would make radar detection quite easy; but since radar frequencies are ultra high frequencies and are therefore useful only in line of sight or straight line operations — that is, the energy waves of radar do not follow the earth's curvature. Additionally, there is a clutter factor close in to the radar antenna which helps to obscure low-altitude targets and hinder their being detected.

On this designated day, six, four-plane divisions of AD Skyraiders were to take off from the aircraft carrier some one hundred fifty miles off the Southern New Jersey Coast and fly to various targets <u>at altitudes below two hundred feet</u>. Each division was assigned a different target, a target to be reached by a specified but circuitous route of roughly seven hundred to eight hundred miles inland. The AD Skyraider, a marvellous work horse of an airplane built by Douglas Aircraft Corporation had a Radial 3350 cubic inch engine Built by the Wright Corporation. It provided up to three thousand horsepower, and enabled the aircraft to cruise at some 170 to 180 knots (194 to 205 miles per hour).

Navy flight "Bravo Three," a four plane division, under the leadership of LT Hank Miller was assigned the mission to make a simulated bombing attack on Wright-Patterson Air Force Base near Dayton, Ohio, without being detected or, if detected, to try to complete the mission under fire. Because of the roundabout route assigned, the distance to the target was close to a thousand nautical miles; and based on a speed of about 175 knots (a bit over 200 miles per hour), the flight time each way was just under five hours or a total flight time of about ten hours. Flight Leader Hank Miller was a good, experienced pilot with 2500 hours in military aircraft and several hundred in the AD Skyraider. The other pilots in his division were experienced and competent as well. Since these flights had been going on for several days with differing targets, the pilots had a good idea of what to expect, although Miller's division felt relatively grateful that they had not been assigned a longer flight of twelve, fourteen or even sixteen hours as some others had. The Air Force had been alerted to expect these attacks and maintained their aircraft in an alert status to jump on the Skyraiders as soon as they were detected, so a circuitous route had to be used. Flight "Bravo Three's" route included the takeoff from the carrier north to cross Long Island near Riverhead, then up to New England north of Montpelier, Vermont, then west and south across New York State and south to Wright-Patterson Air Force Base near Dayton, Ohio. This circuitous route was like flying three sides of a rectangle and would bring the attacking aircraft in from a more northerly or northeasterly direction than a direct line from the Atlantic Ocean off New Jersey. The route back to the carrier was to be somewhat similar.

Although one aircraft carried a radar operator underneath and behind the pilot's cockpit, all the aircraft were single seat aircraft with cockpits that left no room in which to move. So, it takes but little imagination to realize that for the entire ten hours, the pilot was strapped into what would be regarded as somewhat less comfortable than a straight-backed kitchen chair. Although the pilot sat on a small cushion on top of some meager emergency supplies, which were on top of his seat-pack parachute, it would not take long for everyone's posterior to be sensitive to each lump beneath him, the discomfort level increasing as the flight progressed.

To simulate realistic conditions, no electronic navigational aids were to be used. That meant that the navigation was to be a combination of dead reckoning (time and airspeed computations) and visual observations with the help of excellent terrain maps which gave very good depictions of landmarks such as cities, lakes, rivers, railroads, etcetera. Given these limitations, it was essential to fly over intended landmarks as closely as possible to avoid mistaking one landmark for another it resembled. Missing an important landmark could result in an aborted mission. Though a fine aircraft, the AD Skyraider did not have great creature comforts. For one, these AD's did not have an autopilot; not having an autopilot meant that the pilot had to have his hands on the flight controls at all times — though, even if autopilots had been available, how many pilots would be comfortable to have the aircraft being controlled by the autopilot at such a very low altitude is pure conjecture. But on all of these low-level, long-range flights, there were some times when help of some kind of autopilot assistance would have been appreciated.

This mission proceeded very well. And Hank Miller was pleased that they seemed to be hitting their navigational check points in good order. But Bravo Three was soon to realize that conditions change. For one thing, the visibility, which had been about five miles, seemed to drop to about a mile and a half. And though the low altitude enabled them to see the ground clearly, it became difficult to get a good feel or perspective on the surrounding terrain. Mountains were not a problem, but there were many hills which required the pilots to almost constantly adjust their altitude to maintain their assigned two hundred feet or lower altitude. Nevertheless, there were some good landmarks like groups of lakes which were easy to find. Theirs was a mission of maintaining the correct altitude, trying to navigate, and addressing a concern of an essential nature: although these pilots were young men, ten hours is a long time, much too long to reasonably expect a pilot not to relieve his kidneys. Provisions for using a toilet were nonexistent, except for a device known as a relief tube. This was literally a rubber tube vented to the outside. The rubber tube was about one half inch in diameter with a narrow funnel at the end accessible to the pilot. Though not everyone needed relief before their target was reached, everyone would, no doubt, need it at some time during this long flight. The routine of using this device was a busy one, comical enough to provide laughter were a film or video available. One can imagine the goings-on in the cockpit when the pilot would attempt to use this relief tube. Flying at two hundred feet above the ground, rising over uneven terrain that required his adjusting his altitude and throttle constantly, trying to navigate, trying to loosen his flight suit sufficiently and releasing his seat belt a bit, and trying to manage all of this with one hand, and then getting everything back in place, the pilot must have presented a comic scene on this ten hour flight — one can imagine the frustrations that the pilot confronted. A video of all this played back aboard the carrier would have provoked feelings both of laughter and anger. Food and drink have not been mentioned, but they too presented their own problems. For very good reasons, food and drink could not be taken until after the attack and escape had been successfully completed.

Although fuel problems are much more common in jet aircraft because of their high fuel consumption, particularly at low altitudes, they do occur as well in piston engine or reciprocating engined aircraft, the rate of consumption also being affected by altitude. Some would argue that, when you consider the speed with which jets fly, the situations are pretty much the same because they cover more ground in a shorter time. For example, a jet could cover the same distance as a piston engined aircraft in less than half the time, so the jet need not have as many minutes or hours of fuel to cover that distance. Where exhausting fuel at sea would be a problem in piston engined types, most likely the pilot's navigation would be at fault; for example, when a pilot reached an area where he thought the aircraft carrier should be and saw no sign of it and could not contact the carrier by radio for help, he would most likely start an approved search procedure. During this process, he could exhaust all fuel before he located the carrier. Navigational problems over the years have resulted in the loss of quite a few aircraft and pilots. The same type of problem can occur over land.

Having reached each navigational check point exceedingly well — better than they had hoped — they started their run in to make simulated bombing runs on Wright-Patterson Air Force Base, encountering increasingly deteriorating weather, especially decreasing visibility. Somewhat concerned about the possibility of a midair collision with other traffic in the area, though none should be in the area during this exercise, LT Miller called Wright-Patterson to announce the start of their attack. The bad weather helped conceal the attack, making it a very precise attack and a complete success; but the call had alerted the base to the attacking flight, and as Miller's flight of four completed its escape turn out of the area, they found themselves being accompanied by two Air Force jet fighter aircraft. Yes, they had been intercepted, but there was no question about their having successfully accomplished their mission. Had they not announced their position just before making their attack, there was no way they could have been detected on radar at their low altitude, let alone be intercepted. They would have been in and out in a flash. But because of the bad weather, particularly the visibility, the call to Wright-Patterson was a concession to safety.

Now safely away from Wright-Patterson, they still had to maintain an altitude of two hundred feet above the terrain to minimize their being tracked to their point of origin, the aircraft carrier; they then had to navigate north and east to Montpelier, south to Riverhead, Long Island, and then cross over one hundred fifty nautical miles of open sea to a point where they thought they would find the aircraft carrier. The flight, which seemed to pass slowly on the way in to Wright-Patterson Air Force Base, now seemed to pass more quickly as they distanced themselves from Wright Patterson and as they consumed some of their lunch and headed south from Montpelier toward Riverhead some eight hours after having taken off from the carrier at 0700 this morning. For the AD Skyraiders taking part in this one week exercise, three large external fuel tanks consisting of a tank under each wing and one under the fuselage had been attached. With some similar configurations, there could be enough fuel for fourteen, sixteen hours, and perhaps up to eighteen hours of flight. Having literally been strapped to an uncomfortable chair-like seat for the last nine hours and now crossing the south shore of Long Island headed out to sea, their restricted confinement seemed to make the pilots increasingly fatigued. Their heading was one hundred fifty-five degrees; and since the New Jersey Coastline runs more southwest to northeast than south to north, their heading was taking them nearly at right angles to the coastline or directly away from land — just exactly as planned. LT Miller, using the tactical frequency to which his flight was tuned, told LTJG Hal Stark to climb a bit and have his radar operator try to pick up the carrier on his radar scope. Stark's radar man, seated behind and below the pilot in the lower part of the fuselage, was known to be one of the best radar men in the squadron. So, it was with some disappointment to hear that, after trying for ten minutes, he could not get anything at all to appear on his radar scope. He was advised to use his beacon mode, a method of causing the carrier's radar to respond to the aircraft's trigger signal. Such a beacon signal would increase significantly the range at which you could receive information on the radar scope. A blip would appear on the aircraft's scope showing the range and bearing of the ship — provided that both the aircraft's radar was working and the carrier's radar was transmitting. But of course neither situation prevailed, and no information could be obtained. It seemed logical that, due to the nature of the mission, the carrier would not be emitting electronic radiation routinely. Such a condition was not required in this simulated exercise; but nevertheless, the carrier was silent electronically. Even without the confirming information the radar would have provided, Hank Miller felt reasonably sure he was leading his division on the right course to the carrier; but with no backup information to cross-check his position, there was no way he could be certain. Had their one radar set in the division been able to pick up the carrier, or had their beacon triggered a response from the carrier, there would have been no concern at all. What did concern Hank Miller was the state of fatigue of his men nearly ten hours after their morning takeoff from the carrier. After all, anyone so strapped into a cockpit space no bigger than about two feet wide and a couple of feet from the instrument panel could not help but be uncomfortable. Nor could Hank not be concerned that his pilots might be becoming a bit apprehensive about the lack of communication with the carrier.

It seemed to LT Miller that the flight must be within twenty or so minutes of the carrier, but it was not yet in sight. Calling on the tactical frequency, Hank Miller advised his wing men to switch to the aircraft carrier's frequency. Then Hank called the carrier and received no response. He waited a minute and called again — still no response. Another minute and another call — no reply from the carrier. Had they not heard him? That seemed inconceivable. He told his near wingman to call the carrier — still no response. What was the problem? Even with the carrier in sight, radio communication with the ship was necessary for permission to land aboard. Were they so far away that the ship could not hear them? Was the carrier observing complete radio silence? Nothing was mentioned about that during the morning's briefing. What to do now? Hank was not yet really concerned. He was, however, concerned about his wing men's state. Were they growing anxious? He did not think so, but could not be certain. But it was now a little more than ten hours since takeoff, and visibility would not be improving in the approaching dusk, making a visual sighting of the carrier more difficult — and the developing fatigue would not make things easier. Yesterday, a pilot returning from a similar long-duration low level flight had gotten careless in his approach turn to a landing on the carrier and spun into the water, killing his radar man who was trapped in the fuselage. Flying low and slow as is required in the approach turn into the carrier is the worst place to get careless, especially if the pilot is very tired. Everyone felt that fatigue had been a factor in that accident. Keeping in mind the duration of the flights and their enervating nature and attendant risks, one pilot who returned to the carrier around midnight in the midst of a snow squall stated the obvious when he said, "This is a hell of a way to make a living.!" From their present position, Miller figured they were about an hour's flight from the Naval Air Station, Lakehurst, New Jersey, and that they still had about an hour and a half's fuel. Hank Miller wondered just how many minutes more he should pursue his present course before heading for the beach and Lakehurst.

Considering all the factors, Hank Miller felt that the "crunch time" had arrived and that a decision had to be made. Either he was in the location where he thought he was, the general vicinity of the carrier, or he was not. If he was in the vicinity, then the carrier had heard him and, deliberately, for some reason or other, did not intend to reply. If he was not in the vicinity, then he had better lead his flight toward the beach where they could use some electronic means of navigation, rather than to continue a course that possibly led away from both the carrier and the beach. In a form of mild understatement, Hank paused to wonder whether it would be dinner aboard ship in the stateroom or dinner at Lakehurst. Or was it to be no dinner in a chilly ocean. LT Miller knew what he was going to do. If the ship had heard him but not responded, it was probably an indication that his flight was on course or nearly on course. If that was the situation, Miller knew how to get a response from the carrier. Since all four aircraft in Hank's division had been on the carrier's radio frequency since his first attempt to call, he now decided to advise the pilots about what he was going to do — deliberately using the carrier's frequency rather than their tactical frequency, which he was not sure that carrier would hear. In so doing, he was sure the carrier people would hear him. Accordingly, LT Miller announced to his wing men, "I'm going to call the ship once or perhaps twice more to try to reach them for a steer to the ship. If they don't respond, we're heading for Lakehurst. I don't want to put four aircraft in the water!"

Hank had barely begun to consider the precise wording of the call he was going to make, when he was startled by a very loud, very strong, and very clear call from the carrier in anticipation of Hank's imminent call. "Roger Bravo Three, we have you northwest at twenty-two miles. Your 'pigeons to base' (heading to the carrier) 152 degrees." LT Miller made a slight left turn from the 155 degree heading he had maintained throughout the flight to the new 152 degree heading; and in just a minute or two, sure enough, there at the limit of their visibility — about twelve miles — lay the carrier, now a beautiful, inviting creature that would welcome them all to its friendly bosom, give them a nice dinner, provide them with a warm, dry place to sleep — and then send them out in the morning on a similar long distance, low-level flight. But first, aware of their tired condition and mindful of yesterday's fatal crash, each pilot had to be especially careful in making his approach turn into the final approach to the carrier. Each pilot did not have to be reminded that, up to this time, the flight had been a good flight: they had handled the navigation, maintained their low-level flight, dealt with the flight's imperious demands — and they had been able to attack their target without being detected! Even their route to the carrier over the open water had been "on the money." It had been a successful if enervating flight, and now no one was going to spoil this flight by messing up the final couple of minutes during his landing aboard.

Once aboard ship, the pilots climbed out of their aircraft; and even though young and fit, they had to overcome a degree of stiffness from their confined positions. Helping them to unbend and unwind, was a greeting from the flight surgeon, who, hustling them into his flight deck emergency station in the island superstructure, administered his usual treatment to pilots returning from these long flights: three ounce bottles of brandy, a big surprise to the pilots who knew that alcohol aboard ship was strictly forbidden, but who were reminded and could readily see, of course, that this was really medicine. No one felt he had a right to refuse the doctor's orders.

Postmortems never seemed to receive much attention; perhaps this flight deserved no postmortem. But it is only conjecture as to why the carrier had not responded to calls from Bravo Three. The most likely reason is that radio silence was part of the exercise, though certainly that had not been mentioned as part of the morning briefing. Also, Bravo Three had been pretty much on course to the carrier; and the carrier may have wanted to see how things would evolve, perhaps thinking that the ship's position would reveal itself. It seems likely, however, that, had LT Miller's flight been far off course and in danger of passing too far to one side to see the carrier, the carrier would have contacted the flight. Two points remain to be explained. If the ship's radar had been transmitting, why had the air crewman not been able to trigger a beacon reply? Most likely, the aircraft's radar was not working properly; but it is also likely that the carrier's radar may have been in a standby mode — that is, not transmitting continuously but ready to transmit in compliance with radio silence requirements. It is possible that the carrier was transmitting intermittently for short periods of time, enough to keep track periodically of the position of aircraft returning from the day's missions. Since this flight was one of five or six similar flights this day, time considerations did not permit the luxury of a critique long enough to discuss the last minutes of Navy "Bravo Three."

In Search of the Submarine — On a Very Black Night

So far, it had been a good active duty training period. Now, eleven days into the fourteen-day active duty period, Operations Officer Lou Benson calculated that the average flight time per pilot would easily exceed fifty hours — not a bad sum for so short a period of time with so much to do and with each hour of flight to be used to obtain further readiness qualifications. From the first Saturday afternoon in early October when Anti-Submarine Squadron VS-980 had departed from the Naval Air Station Willow Grove, Pennsylvania, and flown to the Naval Air Station Norfolk, Virginia, in their S2F Trackers, things had gone exceptionally well. The flight route from Willow Grove had been first via Woodstown, New Jersey, then Dover, Delaware, then Patuxent River, Maryland, and Richmond, Virginia, and on to Hopewell, Virginia, north and west of Norfolk, where Navy Norfolk Approach Control picked up each plane as it reached this point and guided it in the deteriorating weather to Navy Norfolk via a radar Ground Controlled Approach (GCA).

As the squadron's operations officer, Benson was in charge of training and qualifications, that is readiness, and had planned well to insure that each pilot was scheduled for and flew the requisite number of flight hours in the areas required to reach the highest state of pilot and crew readiness. Besides the routine hours of instrument flying (real and simulated bad weather) and night flying, there were flights of rocket firing and bombing, and some flights over the submarines for tactics, etcetera.

On this eleventh day, the most important flight remaining was an operational readiness flight, a flight to demonstrate to impartial observers from outside that the pilots of VS-980 could apply their skills effectively in a night mission against a U.S. submarine located about one hundred miles east of Virginia's Cape Henry. The mission would have four requirements or parts: first would be locating the submarine which could either be on the surface of the water or, if not on the surface, "snorkeling," a configuration which allowed the submarine to operate just below the surface of the water yet with its air intake valve or "Snorkel" remaining just above the surface of the water, insuring a supply of air for the sub's diesel engines. Operating in this mode, the submarine could run on its oil burning diesel engines instead of using its precious batteries, which are required for operating when the submarine is completely submerged. With just the Snorkel above the surface, the submarine presents a very small area to reflect an aircraft's radar transmissions, making it much more difficult to detect. The observers would be from an active duty squadron based at Norfolk and would be unknown to the Willow Grove pilots assigned to this operational readiness mission. In fact, the pilots from each squadron never did meet each other, not before takeoff and not afterwards: the Norfolk observer pilots were to join Willow Grove pilots after takeoff, and later would submit their reports. An essential part of this report would be an overall grade for the degree and the manner of completion of all the tasks the observer pilots would assign. All of this was, of course, contingent upon the mission pilots' being able to find and operate with a submarine with such an unspecific location — and on such a very black night as this night would become.

The second part of the mission required backing away from the submarine and then relocating it using passive electronic means. That meant there could be no electronic transmissions like radar, but rather only electronic "listening" using electronic countermeasures or "ECM" as they are usually referred to. This countermeasure required the aircraft to pick up or receive the radar transmissions of the submarine and, by using a simple geometric triangulation solution, estimate the submarine's distance from the aircraft.

The third part required the use of radar or "active" measure transmissions to relocate the submarine, determine its distance, and then close in or "home in" on the submarine and drop a weapon. Before the weapon could be dropped, the procedure required the aircraft to make a gradual descent to about two hundred feet as it approached close-in to the submarine, at which time the copilot would turn on the seventy million (70,000,000) candle power searchlight and light up the submarine for positive identification before dropping the weapon.

The fourth part, and the most dangerous, would require the use of magnetic anomaly detection equipment, or "MAD gear" as it is usually called to pinpoint the location of a submerged submarine. Use of such equipment would be required when the submarine had completely submerged during the aircraft's run in on the submarine. For example, the submarine might detect the aircraft's radar through its own radar or ECM equipment and attempt to dive as quickly as possible. In this situation, the aircraft would be left with an approximate position of the submarine, a position not accurate enough for a weapon drop. So the use of MAD gear would be employed to detect simple variations in the earth's magnetic lines of force, variations caused by the presence of the submerged submarine or other metallic substances under the water. Although the equipment sounds wonderful — and it is — it has very limited range, requiring the pilot to fly very close to the water and nearly directly on top of the submarine to maximize detection possibilities. Difficult in daylight, it becomes much more so at night and even more so on a very dark, horizonless, overcast night when the pilot is really flying on instruments and must rely on his altimeter when very close to the water — and altimeters can be less than completely reliable for a variety of reasons. It is very difficult for pilots to judge their altitude over the water in daylight, particularly when there are no waves. At night it becomes much more so.

In considering whom to schedule for this important mission, LCDR Benson felt surely that the Skipper, Hal Wange, would want to be included. Benson also felt that three aircraft could do the job, allowing insurance for the possibility that perhaps one aircraft would have a mechanical problem. With three aircraft, it was felt that actually locating the submarine would be easier — an important factor since, without the submarine, there could be no mission accomplishment. With the Skipper leading the flight, Benson chose the Executive Officer, LCDR George Malson and crew for the second aircraft; and Benson wanted his crew to be there as well, flying wing on the others. Benson wanted to see first hand how things would go and how well some of the young crewmen would perform. Each crew consisted of a more junior officer as copilot; and then there was a radar man and an ECM operator. There was much pride and competition among the crews, with each pilot claiming he had the best crew. Takeoff time was scheduled for 2000 hours (8:00 PM), well after dark at the end of the third week of October; and the briefing was set for 1830 hours. All four members of each crew were to attend the briefing, so that everyone would have knowledge of the complete mission.

By 1815 hours and not wanting to keep the skipper waiting, two crews had put on their flight gear and were assembled in the ready room awaiting the arrival of the Skipper and his crew. "Strange," thought Benson as he looked around the ready room. But there it was for all to see, only no one had noticed it: The Skipper had drawn a line through his name and initialed it, removing himself and his crew from the mission. Benson was very much surprised by this development; but Benson felt that CDR Wange must have had good reason for the change, even though he could not help but have some private thoughts about a change so late that a replacement could not be obtained. But there was no time to dwell on the "why's" at this time. There were still two very good crews for the mission, enough for a good mission accomplishment; and besides they might be better off doing it on their own. It was not that they did not welcome the Skipper's leading the flight. It was just that Benson and Malson and their crews knew each other well and had great confidence in each other. So the flight, though demanding, might be conducted with a little less tension without the Skipper. George Malson was a solid, serious, capable, and reliable pilot; and he conducted the briefing in a businesslike manner going over the weather, tactical frequencies, and the call signs (radio code names) of the aircraft, the observation plane and also the submarine. The VS-980 Flight would be "Diamond Flight," its leader being Diamond One and Benson's aircraft being Diamond Two; the observer aircraft would be "Revolver One," and the submarine was to be "Sunshine," a strange name considering it usually saw but little sunshine. Malson spent some time explaining that, first and foremost, they had to find the submarine some one hundred miles east of Cape Henry, which is about fifteen miles east of Naval Air Station Norfolk. They knew this task would not be easy since they could use no electronic navigational equipment. Even if permitted, the only electronic equipment they could use would be TACAN, equipment which could give them bearing and distance from Navy Norfolk. And it would be very doubtful that they could receive the TACAN signals at the altitude they would be flying and the distance of the submarine from the equipment. In any event, they would not consider using TACAN when the mission requirements prohibited it. So, locating the submarine would be difficult on this dark, moonless, starless, overcast night. Location would have to be primarily by the aircraft's radar; or if the submarine's radar was transmitting, perhaps the ECM operator could get a bearing on the submarine by zeroing in on those radar pulses. At best, the aircraft's radar man could expect to get only get a pinpoint blip on his radar scope. If there were ships out there, their blips, being larger, could mask the submarine's blip. Also, if the sea was rough, the radar would pick up "clutter" caused by waves, perhaps hiding the blip. Fortunately, the sea state was forecast to be calm and the air free of turbulence. Malson then explained that, most likely, the observer pilot would assign the first mission to one crew, while the other crew orbited at a distance; and then the second crew the next, then back to the first crew and so on. There would be no way of knowing what part of the mission anyone would be asked to handle. Such an assignment of parts would be a way of observing whether both crews were proficient in handling all of the parts. Although Benson felt that both crews could handle all parts equally well, provided they could find the submarine in the first place, he could not help but worry that both would be sure to exercise the proper care in its required operations so close to the water on such a dark, horizonless night. (All of aviation, military and civilian has its stories of aircraft unexpectedly flying into the ground or water.)

The briefing now ended, it was time to get out to the aircraft for the 2000 hours takeoff. The pilots would have to start their engines, take off, and rendezvous in a loose formation. Right after takeoff, the pilots would switch to the tactical frequency for mission communications. Malson was the first aircraft to takeoff, followed immediately by Benson, and then the observer pilot. As Benson joined up on Malson and switched to the tactical frequency and called to check communications, he had a good feeling about this flight. To Benson, Malson was as good as they come, being completely dependable and capable. So, much to his surprise, Malson returned Benson's call with the message that he did not like some of the engine instrument readings he saw and that good judgment dictated that he return to base. Benson knew that, if Malson thought he ought to return to base, he must have a very, very good reason. As one of the best, Malson was not one to get excited about little things. Now Benson was seeing the entire flight fall apart — and a flight of considerable importance. In all of this active duty period, very few aircraft failed to get off the ground or had to return to base after takeoff. Every mission had been completed. Was this just bad luck? Now Benson wondered why the Skipper did not assign someone else in his place. Perhaps it was the fact that things had gone so well during the entire period. And perhaps he felt so sure about Malson and Benson. Nevertheless, the Skipper was experienced enough to know that you just cannot predict everything that will happen and that a little insurance is not a bad idea. At this point, Benson was a bit annoyed with the Skipper's leadership.

It seemed pointless now to concern himself about these things that had happened so far, when a decision to proceed with the mission or return to base had to be made. As Lou Benson knew, Revolver One had heard these transmissions on the tactical frequency and was waiting for some word from Benson, who now became Diamond One. Speaking to his copilot LT Joe Robby, Lou said, "Let's go Joe and hope for the best. Benson knew that he had a good gutsy, competent crew, the best in the squadron: Robby was a good copilot, confident and possessing a good sense of humor; Radarman Roger Black could get results out of any radar that could be turned on; and ECM operator Al Pflaum was a solid, reliable man who knew his business and could be depended upon to produce results. The crew's decision to continue the mission now made, Benson called the observer aircraft: "Revolver One, this is Diamond One. Proceeding on course to the operating area." Revolver One acknowledged with a simple, "Roger," and the flight was under way. Now, with but one aircraft, Benson's only misgiving concerned whether they could find the submarine, a speck in a great expanse of water. But additionally, he hoped all of his aircraft's equipment would work properly. He thought that a poor job now on top of the other problems encountered so far would be a real fiasco. Would it really have been better to return to base and try to reschedule another operational readiness inspection? At best, that would be Friday. And what if they wouldn't reschedule it, feeling that the squadron had had its chance and failed. The Skipper and Exec would have to give an accounting. Lou Benson put these thoughts out of his mind in an instant, feeling that he had to get on with the job at hand and do what he could. If anything, Benson and his crew were more determined that they give a good account of themselves. Besides, Benson knew that, if his own aircraft had developed a problem, Malson would have gone on alone and done a good job — perhaps better than anyone in the squadron.

Well out to sea by now, Benson and Robby had had to maintain a steady course to compute their estimated arrival time over the submarine using the wind force and direction they had received, and to estimate at what distance they could reasonably expect to pick up the submarine on radar. Since no electronic navigational equipment had been used, Benson was very careful to maintain his desired heading, feeling that, if winds were a little different than forecast, they probably would not be too far off their intended course. The very black night prevented their seeing any waves on the water which could have indicated wind direction and, less accurately, wind force. But even if they could have seen the water and obtained surface wind information, it surely would have been different at the altitude at which they were flying. At best, they might get some information whether to expect to be a little left or a little right of course. With the submarine now estimated to be about twenty-five miles distant, LCDR Benson, with some hopeful anticipation, asked Radarman Roger Black, "Have you got anything yet, Roger?" Roger's reply was not a crisp "Yes" or a crisp "No," but something not entirely unanticipated: "Well, I think I might have something at about twenty-two miles, slightly to our right. But I'm not sure; I'll try to sharpen up the image." It was the best Benson could hope for, knowing that Black was sharp and cautious and not the kind to say "yes" when he was not certain and have the pilot chase the wrong target. Not many seconds later, Roger reported a very tiny, but very clear blip, explaining that this very well could be the submarine since it was too small to be a ship. LT Robby adjusted the repeater radar scope in the cockpit, and a mere pinpoint of light, the target, came into focus. Still not absolutely sure that it was the submarine despite the very encouraging signs so far, the uncertainty was removed when ECM operator Pflaum reported, "I'm receiving radar signals on the ECM gear in the proper frequency range, bearing fifteen degrees relative, or fifteen degrees off their heading. Benson felt proud of his crew and, at the same time, much relieved. Had they missed picking up the submarine, there would have been no mission accomplishment at all. Had Benson been overly concerned? Was there but little likelihood of their missing the sub? Perhaps. But Benson had been around for awhile and knew some of the horror stories concerning aircraft that had missed ships, usually aircraft carriers. Whatever would be asked of them now, Benson felt sure they could handle it. Yet, with all of the electronic information gathered so far, there had been no visual sighting of the submarine — and there would not be for awhile.

Certain that they were approaching the submarine, Benson picked up the microphone, "Revolver One from Diamond One, have submarine 015 degrees relative at ten miles; please advise. Revolver One's instructions were brief and crisp: "Diamond One, turn radar "off" and take heading 245 degrees." This instruction was followed at three or four minute intervals with new heading instructions, all taking them farther from the submarine. At this point, Revolver One asked Diamond One to conduct several triangulation exercises using ECM equipment to determine the new bearing and distance to the submarine. Joe Robby "Rogered" Revolver One's instructions for Diamond One, and Benson then asked ECM operator Pflaum to take over. Pflaum was not long in reporting the submarine's bearing as 135 and asked the pilot to take a heading of 225 degrees, a heading ninety degrees to the submarine's bearing. After three minutes, Pflaum reported the submarine's bearing now to be 125 degrees and that, using the standard triangulation method, the submarine was roughly seven minutes or twenty-one nautical miles distant. Prudent procedures required that, unless there were urgent reasons, the first determination be checked by a second; so Benson instructed Pflaum to make another check to verify the first finding. Three minutes after turning to the new heading Pflaum requested, Pflaum reported the same distance but a slightly new bearing. Joe Robby called Revolver One: "Revolver One, we have 'Sunshine' (the sub) at 105 degrees, twenty-one nautical miles." Not a very precise method of submarine localization, the use of ECM equipment is, nevertheless, a reliable passive procedure, useful either when the aircraft's radar is not working or the pilot does not want to use his radar for fear of being detected by the submarine. Revolver One's response to Robby's report of Sunshine was prompt in coming: "Diamond One, turn radar 'on'; verify ECM's bearing and distance, and commence low level simulated bombing run using radar for homing and the searchlight to light up Sunshine." Robby Rogered Revolver One's instructions, reported the location of the submarine's position as confirming the ECM's location, and reported the commencement of the run-in to Sunshine. Now at fifteen hundred feet and about twenty miles distant, the actual beginning of their radar run, Benson alerted Robby to put the seventy million candle power searchlight on "standby" position. The run-in procedure called for a systematic letting down as the aircraft got closer to the submarine, so that, by the time they were one mile from the submarine, they would be at about two hundred feet of altitude. At this point, they would light up the submarine with the searchlight and, assuming they had the right target, make a weapon drop. Prudence required that this maneuver be exercised with good judgment, particularly so far as altitude is concerned. When the pilot could not be exactly certain that his altimeter was set for atmospheric conditions so far from home base and when the night was so black, inattention to detail could bring fatal consequences. With no horizon and not being able to see the water below, they had no visual references and so were flying on instruments alone. Radarman Black reported the submarine was now ten miles distant, and Benson felt that their eight hundred feet of altitude was just about right, as they headed toward the tiny pinpoint of light on their cockpit repeat radar scope. From here into the submarine, they would really have to concentrate on the flying and on the procedures and leave the distance reporting to Black. Both Benson and Robby, who would turn on the searchlight at the precise time to light up the water and the submarine, became ever more aware of how very black this night was: There was no horizon, no up or down, absolutely no sense of motion in the calm, stable air mass; there was only their complete immersion in a black environment void of any sensation. They could have been in a dark room or closet looking at some dials or instruments. But this was reality, and not a dream, and these were very important instruments, their lifeblood so to speak, since, in this black milieu so dangerously close to the water, water they could not see, they depended on them completely for attitude, altitude, rate of descent, engine operation, and everything else.

Now, passing through five hundred feet and approaching the five mile point, they could see neither the water close below nor anything else — it seemed to be a black fog they were flying through. They were flying as much on instruments now as they had ever flown in the most severe weather conditions. They knew the difference, of course, because it was not only when flying through clouds or fog that flying on instruments was necessary; there were many different kinds of conditions requiring instrument flying. They knew also that they were getting close to a critical point where some pilots, either through overconfidence, lapses in judgment, or plain carelessness, have flown into the water — and both Robby and Benson were aware of this, each pilot checking the other for attention to detail and precise procedures. After all, they well knew that the goal was to sink the sub, not destroy the aircraft and kill the crew. Radarman Black was now calling off each mile; and at two miles and three hundred feet, Robby, anxious to use his searchlight, counted off ten seconds and lit up the water just below as Black called out, "One mile!" Turning the searchlight from standby to the on position, Robby swept his powerful light on the water slightly from left to right and then ahead, when, as expected, the submarine came fully into view. Lit up from bow to stern, it was a beautiful sight, confirming their expectations and making a simulated bomb drop a piece of cake. But until they lit up the submarine, they could not be absolutely certain that they had the sub and not some fishing boat or even a large buoy.

They passed over the sub at one hundred fifty feet in an instant, Benson quickly checking his altimeter for accuracy, and they started their climb to a more suitable altitude. As they climbed to make another radar run, Benson did some calculating and reflecting. The flight to the operating area, the ECM procedures, and their simulated bombing run on the submarine had required about one hour twenty minutes. It had occurred to Benson that, although all signs indicated they had been working with the right submarine, they had not actually communicated with it (That was the observer's job.). They had talked only with Revolver One. How humbling, even humiliating and worse, it would have been, even if understandable, if they had lit up and bombed the wrong submarine.

Revolver One instructed Diamond One to make two more simulated bombing runs utilizing radar homing, advising Diamond One that during the third run, he was to assume the submarine would submerge just before they lit up the searchlight or about two miles before they reached the sub's position. Submarines will submerge if they can, as a means of avoiding just such a bombing run, so this would simulate a realistic situation. This different situation on the third run called for the use of the special airborne equipment to detect the accurate position of submerged submarines. This equipment is, of course, the magnetic anomaly detector or MAD mentioned earlier. Satisfied that his altimeter was set reasonably accurately, Diamond One made the second radar bombing run as accurately as the first; so they approached the third run and the add-on MAD detection mission in a confident manner, determined to concentrate on all aspects of the exercise to do the job both well and safely. In a wartime scenario, the pilot would want to drop his homing weapon in the quickest manner. If he visually observed a hostile submarine going beneath the ocean's surface, and he then received a detection signal on his MAD equipment, he would deem it logical to drop the weapon on the first MAD indication, feeling that that might be his best chance of getting the submarine. In practice exercises, however, two successive MAD signals have to be received before the weapon (simulated) would be warranted. On this night, since they were permitted to use the searchlight and since the submarine's general presence was known, two signals would be required in addition to the signal the MAD operator would get during the radar homing run.

MAD operator Pflaum had turned on his MAD gear and was ready as the aircraft closed in on the still submerging submarine during the third radar run. As they passed over the lit-up but disappearing submarine on the third radar run, Pflaum called out "MADMAN" on the intercom to let the pilot know immediately when he received the detection signal, and LT Robby called Revolver One to advise him that they had received a "MADMAN." Tactics then required an immediate, abrupt, sharp, climbing left turn to about five hundred feet, which would continue through two hundred seventy degrees; during the last ninety or so degrees the aircraft would be descending so that, when the aircraft reapproached the estimated position of the submarine, they would be down to just below two hundred feet of altitude. The idea was to get back to the submarine's position as soon as possible — and at the lowest permissible altitude so that the MAD gear would have the best detection possibilities. Only moments after completing the two hundred seventy degree turn and at an altitude of about two hundred feet, Pflaum shouted "MADMAN" over the intercom, this time over an unlighted or submerged submarine. The powerful searchlight could only be used for thirty seconds or less without the serious risk of burning it out, and the steep turn was so fast that reusing the searchlight so soon again was not really plausible on this first MAD pattern turn. Benson repeated his climbing turn, thinking midway through, "One more signal and we've got our drop." As they were about to drop down again during the last part of the turn, anticipation mounted because the turns required by the procedure are very steep and very vigorous; and the pilot cannot be sure that his turn will put him in the precise spot where he would choose to be. Nevertheless, a third call of "MADMAN" came over the intercom; and Robby, not knowing whether the submerged submarine could hear him, advised Revolver One, "Second MADMAN, weapon away!" He further reported that they would continue the patterns for two more MADMAN signals and drop another weapon. This would complete the full four turns in the clover leaf pattern. Quite unexpectedly and not knowing the submarine had surfaced again, they received a call from Revolver One: "Diamond One, negative! No more turns! Good show! Let's go home!" Though the flight had not been a long flight, being a little over two hours so far, Benson felt it would likely be another forty minutes to base and then to taxi in to the flight line. Only then could they relax somewhat. Considering that they had accomplished a few things and that the hour would be late, Benson felt that Revolver One's decision was probably a good one. After all, by the time they had their debriefing, had taken off their flight suits and dressed in their uniforms again, another forty or so minutes would have elapsed. It would then be past 2330 hours. Altogether, including preflight briefing time, over five and one half hours would have been required for the flight — and this after a busy day and an early tomorrow. Considering everything, Benson felt Revolver One's saying, "Let's go home" was right on the mark.

This night's being dark, quite black in fact, the air still continued to remain very stable with absolutely no turbulence of any kind. So the flight back to base was a continuation of the sensation that there was no motion — just the same dark room with its dimly lighted dials and a background of the quiet, muffled machinery noises caused by the engines. On such a night, one would expect the return to base to be routine; but to Benson's mind, there was no such thing as a routine flight. Copilot Robby served as navigator, while Roger Black, as usual, was doing his navigation with the radar and had already picked up the outline of the Virginia Coast on his radar scope — and to the north the big opening to the Chesapeake Bay and Hampton Roads. Benson kept his eye on everything, his eyes constantly scanning the gauges for aberrations in temperature, pressure, and voltage readings, which might give some indications of impending engine or generator difficulties; but what he saw was a perfectly normal situation with all of the instrument pointers right where they should be. Benson's attention to these matters was not born of apprehension or nervousness, but a relaxed appreciation that it was just good practice; it was an appreciation gathered from years of flying, from some emergencies he had encountered, and from reports of difficulties which other pilots had experienced, which the Navy periodically publishes in information to all pilots. And with Benson's attention to detail, he could not help but reflect on the last few hours and what it would mean in the way of an evaluation for the squadron's state of readiness, the squadron's ability to carry out its assigned mission. He was of two minds. In one sense, he felt that the exercises tonight had gone well, even better than he had hoped; in another sense, there were many "but ifs." Of the three crews originally scheduled for this inspection, only two became airborne — and one of those had to turn back for mechanical problems. And although Benson knew that most, if not all, of the crews could have performed as well as he had, what assumptions would the observers make of the abilities of the other pilots in the squadron? Would it be reasonable for the observers to judge that the squadron had a high state of readiness? It was possible, but it did not seem likely. It seemed likely to Benson that the grades for proficiency in the exercises actually performed would be quite good, that the grades for the number of participants had to be low, and that the overall grade, therefore, could not be good. He consoled himself by thinking that his crew had done exceptionally well and that, whatever the grade, he was very pleased and proud of them. And besides, they had gotten the operational readiness flight out of the way and could now use Friday and Saturday to focus attention on all the crews which still had qualifications to complete. Benson's thoughts were interrupted by Roger Black's report on the intercom, "Our feet are dry." It was a standard call to report that, according to his radar, they had just crossed the coastline and were now over land. In this case the report, while standard, was not necessary because the pilots had been able to see the lights along the shore and inland a bit. But the report signaled Black's last report and was the indication that, unless requested further, he was securing or turning off the radar for this flight. LT Robby called Navy Norfolk for landing instructions and clearance to land; and in a very few minutes they were on the ground and taxiing toward the flight line.

Shortly afterwards Benson, Robby, Black and Pflaum were back in the ready room waiting for some debriefing comments from the observers. The squadron duty officer, who had been on the phone when the crew came into the ready room, reported that he had received a call from the observers just moments ago to the effect that no debriefing was considered necessary, but that a report would be delivered to the squadron commanding officer, CDR Wange, in the morning. The crew, still in their flight gear and relaxed now that they had a few minutes to unwind, talked about the flight, all feeling that their equipment had worked well and that it enabled them to prove that they could do the job. They were not patting themselves on the back, just reflecting a quiet confidence and the feeling that this night everything seemed to go well. They all knew that, despite training and ability, things can and do go wrong: equipment can malfunction or be marginally effective; the sea state can mask radar detection of the submarine; and the weather can be so rough as to be nauseating, making every effort that much more difficult. All of these, and other factors as well, can help to frustrate a crew and even make it seem inept. They knew very well also that you were only as good as your last exercise.

The crew got out of their flight gear and changed into their Navy uniforms. Everyone was relaxed, the ideal time to do something together, to unwind and relax before they tried to sleep. Ideally, having a beer or two would have been a great way to further strengthen the bonds that make a good crew. There was nothing that Benson, Robby, and for that matter Black and Pflaum as well, would have enjoyed more, because it would have been an ideal time to stress the concept of the interdependence of each member of the crew, including coordination and team spirit. But, unfortunately, they were running out of time for such a celebration. It was already quite late, there was no suitable place on the base, and they all had to get some sleep and get up early for another flight Friday morning.

After saying "Good night" to the crew, Benson showered and got into bed, his mind very much on the events of this evening. Often after some flights, the mind races. There was much to think about, many unanswered questions. He was certain there would be some discussion about the flight in the morning. After all, the Skipper would surely have some questions about it; and he might already have received the report and grade of the observers. It was his squadron; the readiness score had to be important to him. But all that would have to wait till morning, till after Benson and his crew returned from a rocket-firing flight.

Next morning, expecting questions from the Skipper about last night's flight, Benson received none. Neither were there questions from the executive officer. Benson wondered, "Should I ask the Skipper if he wants a report or should I just approach the Skipper and tell him I wanted to report on our operational readiness inspection flight?" Benson felt he had an obligation to tell him something; but every time he had a couple of minutes, the Skipper was tied up with people or he was not in his office. So passed Friday and so passed Saturday and, in fact, at no time did the subject ever come up. The only reference to the readiness flight came up back in Willow Grove Sunday afternoon, when the entire active duty period was reviewed. The Skipper read from a memorandum sent by the observers mentioning that the squadron had received a grade of <u>eighty-three</u> on a scale of <u>one hundred</u> for the operational readiness flight. Nothing more was said. With so little said or enquired-about the flight, Benson was determined that he himself would not mention the flight again. If he was disappointed, he did not show it, but it did seem a bit strange. He and his crew had done what he considered a pretty fair job and were convinced that the only way the score could have been higher would have been if one or two more of the squadron's aircraft and crews had participated. If LCDR Benson had had any doubts, it was his wondering whether his judgment to go alone and not chance a rescheduling of the mission the next day when more aircraft could be involved would have been better. But he had had to make a decision; and he made it, feeling reasonably certain, that the completion of an assigned mission on the day scheduled was a better choice than hoping for a second chance.

The Agony of the Longest Night

They had to be airborne at midnight! But it wasn't as if they had to take off exactly at midnight, because there was a one half hour window that permitted them to take off in the period from 2345 hours until 0015 hours. This somewhat strange-sounding requirement was imposed by the exercise in progress. A group or pod of submarines was to transit from the South, possibly from Norfolk, Virginia, to New London, Connecticut, and the Navy considered their passage an excellent opportunity to send out their long range SP2E Neptune antisubmarine aircraft to detect their presence, using their electronics and ordnance equipment (sonobuoys) and employing their tactical procedures. Any opportunity to detect and search out submarines was considered worthwhile. A VP squadron, temporarily located at the Naval Air Station, Willow Grove, Pennsylvania, was assigned to participate. Since the submarines were to transit at a distance quite far from the coast, flights of ten hours duration were considered necessary. At the beginning of each flight, some allowance was made for each aircraft to get airborne and climb to altitude; and some allowance for the necessary time en route to the operating area and for the relief of the aircraft on station was also provided. An equal amount of time was allowed for being relieved on station and for returning to base and landing, permitting each aircraft about seven hours on station in the operating area.

While a more detailed description of the SP2E aircraft appears in another episode in this book, for now it is enough to say that, ungainly looking on the ground, it appears beautiful in the air, at least partially because of its "Davis Wing," a straight, symmetrical, gracefully designed shape. An ideal airplane for long range patrol and antisubmarine work, it has a one hundred and five foot wingspan and ninety foot long fuselage. Lengthwise, it has plenty of room for the twelve crew positions required for its type of antisubmarine work; but heightwise, because of its bomb bays, fuel tanks, and other requirements, no one could stand fully upright. An additional hindrance to moving about was that the wing or its smoothly covered spars came right through and across the fuselage, creating a hump which had to be crawled over to go forward or aft. It has four engines, including two jet engines for takeoffs, landings, and emergencies; once airborne, it can operate fairly efficiently and economically to save fuel. In the usual configuration, up to 4400 hundred gallons of fuel, including three hundred fifty in each wingtip tank, enable the aircraft to remain airborne for many, many hours.

The departure time for the submarines was not precisely known, but it was expected to be within the two-day period of this exercise. What was likewise unknown was what operational movements and exercises the submarines would be conducting: How much time would they be operating on the surface of the ocean and how much time beneath? And the ocean spaces are huge. Seven individual flights would be required to cover the forty-eight hours on station, starting at 1200 on Monday and extending to 1200 hours on Wednesday.

The skipper of the squadron at Willow Grove, Commander Jeffrey Simpson, wanted his squadron's part in the exercise to go well, exceptionally well; accordingly he worked tirelessly to make sure that each crew knew everything that was required and expected of them. He particularly emphasized the timing that was required of each crew. Extensive briefing would begin three hours before flight time: There was information to be obtained about the types and routes of the submarines; information to learn about the location of the assigned operating area; information on the radio and radar frequencies to be used and reports to be made back to base radio; information on hostile submarine sightings or detections; and, of course, weather information and many, many other things. The briefing was expected to take two hours, leaving one hour for the twelve-man crew to get out to the airplane, complete the necessary preflight checks, crank up the engines, and taxi out to the runway for takeoff. The timing would allow a little time for the unforeseen.

A good skipper, Commander Simpson expected a lot from his pilots without being a martinet. Strong and serious, yet interested in each of his men, he believed an important part of leadership was providing a good example and being out there in front to point the way. It was much easier for him to get after the pilots, if necessary, if he demonstrated by example what was expected. His thoughts about leadership were confirmed by the willingness of his pilots to follow. In keeping with this line of reasoning, Simpson conferred with his operations officer and decided that he, the skipper, would take the midnight to 1000 hours flight on the first day. He felt that the "midnight shift" would not normally be preferred by others. He also felt that the hours he chose, fairly early in the submarines' two-day transiting period, might be a likely time when his crew could expect some action; also, he had known of operations which were cut short of the full scheduled time. So, in this decision, he was, perhaps, somewhat selfish.

To get the operation off to a good start, Simpson decided to get his first aircraft on station at 11:30 AM on Monday, or one half hour earlier than necessary. That meant that the first crew would have an 0700 briefing for a 1000 takeoff, the second aircraft a 1400 PM briefing for a 1700 takeoff, and the third, Simpson's, a 2100 briefing for a midnight takeoff. Sensing the importance of a good beginning, Simpson went over hundreds of details in his mind, trying not to overlook any item that might be important. He sat in on the 0700 briefing and also the second at 1400, in between taking care of the many things that keep a squadron commander busy. After observing the second briefing at 1400 and watching that crew take off at 1700, Simpson felt that he should try to get some rest before his crew's briefing at 2100.

Getting a light snack for dinner, Simpson then went to his room in the BOQ and lay down for what he hoped would be about a two and one half hours' sleep. That was just the beginning of his miscalculations, for his mind kept racing due to the stimulus generated by the mission and getting things off to a proper start. He told himself that he'd be lucky to get one good hour's sleep and told himself that, even if he didn't fall asleep, being flat on his back was quite restful and almost as good as being asleep. He worried also that his alarm clock might not ring at the time set and perhaps even the squadron duty officer would not ring him promptly. "Silly thoughts," he admitted to himself; but, nevertheless, they were a factor in his not falling asleep. He was, however, comforted in the realization that the planning and execution of his squadron's efforts went smoothly so far. But this realization could not calm him down sufficiently enough to fall asleep. He watched his alarm clock as it neared 2030 and then sounded its alarm, almost simultaneously with the telephone ring from the squadron duty officer.

At 2100, Simpson's entire crew was assembled and ready for the briefing, "all bright and bushy-tailed" as the saying goes when one is alert and ready. Thinking about his crew, Simpson knew he had good men: His copilot LT Carlson was an expert in antisubmarine tactics and the tactical communications required in antisubmarine work; his plane captain, Aviation Machinist Mate First Class Hal Weiss was as good as they come and could be depended upon to scan the engine instruments and watch the fuel flow and engine analyzers; the other crew members were good, well-experienced men and consisted of a radar operator, an ECM or electronics countermeasures man, a passive sonobuoy man, an active sonobuoy man, a MAD or magnetic anomaly detector man, an ordnance man, a radio specialist for long range communications, and, overseeing and coordinating all of the operations of these non-pilot crew specialists, was the experienced tactical coordinator, LCDR Gordon, who would decide what antisubmarine procedures should be used and when they should be employed, subject, of course, to the skipper's veto. The tactical coordinator, or TACCO as he was familiarly called, would take the information of the various crew members, plot it as necessary on the radar screen, and decide the next tactic. Looking at his crew, the skipper thought that these young people looked very much like any other group of good young men. But they were not: What made them different was the skill, the discipline, and the expertise that each had — to say nothing of the courage they had that enabled them to be flying such a long flight several hundred miles out over the ocean at night and in possibly undesirable weather conditions. Simpson did not have time to reflect further on these men, because his thoughts were interrupted by the sound of the opening words of the briefing people.

The details of the briefing took the entire two hours allotted; and the crew, now in their flight suits, proceeded to the Neptune aircraft. Each man made his required preflight inspections, and the two main reciprocating engines were started. At 2330 hours, the aircraft started to taxi to the assigned runway area, where an extensive engine run-up and check-off list would be made. With everything checking out all right, the two jet engines would be started and, when they were sure everything was as they wanted, clearance for takeoff would be requested. After receiving clearance, Simpson pushed the throttles on the reciprocating engines up to forty-eight inches of manifold pressure, the maximum allowable with the brakes set, and then advanced the jet engines to 100% of RPM. As soon as the jets reached 100% of RPM, Simpson released the brakes and advanced his reciprocating engine throttles to fifty-six inches of manifold pressure, the takeoff setting, while his copilot placed his left hand behind the throttles to insure that they did not creep backward as they started down the runway. With four engines at takeoff power, there was the equivalent of about twelve thousand horsepower being put to use. When all that power was available, as it was for normal takeoff, the acceleration was quite rapid and could be felt almost immediately, lift off speed being obtained quickly. There is a point during takeoff that is critical: when there is a loss of power on an engine before safe flying speed is reached. For the Neptune with its four engines, once lift off was achieved and the pilot let his speed build up a little, the loss of one engine presented no hazard, only an inconvenience. But Simpson had never experienced any kind of engine problems in the Neptune, despite having over seven hundred hours in it; and he was not to have any difficulties now. They were airborne and en route to the operating area at 2359 hours.

The night was dark, very dark. A high, thick cloud cover obscured the moon and the stars; and visibility, while not bad at five miles, wasn't enough to make the atmosphere lighter. Simpson realized rather quickly that the air was calm; and once over the ocean, the air seemed completely calm. While Simpson had not previously thought about it and did not yet realize it, the calm conditions were to present problems later. There were two other things which, combined with the calm air, were to make this a night of misery, a night that seemed the longest of his life. On approaching the "on-station" position, Copilot Carlson contacted the crew being relieved for information they had on the location of any submarine contacts they had been working with. To the surprise of all, they were advised that there had been "No contacts!"

But Simpson's crew was charged up. They were ready and confident that they would detect any submarine transiting through the area. After all, they had good equipment, and they were well trained and experienced. Blips would appear on the radar scope, but they were always too big to be submarines on the surface, let alone a submerged submarine with just its snorkel or diesel engine air intake protruding above the water. No radar pulses from the submarine were picked up on the ECM equipment, nor did a line of active or pinging sonobuoys dropped in the water reveal anything like a submarine. After being on-station for an hour and not detecting anything that might possibly be a submarine, everyone in the crew was disappointed with the lack of action; but they were not discouraged, remaining very alert and eager for a contact. In the meantime, Simpson was fascinated with how very calm the air was, thinking he could not remember any conditions quite similar. The noise of the two main or reciprocating engines (as usual, the jets had been turned off shortly after takeoff) was muffled; and the engine vibrations seemed hardly noticeable. In the blackness of this night, it was impossible to tell up from down, or left from right; so it was necessary to fly "on instruments" to insure level, turn-free flight. In such conditions and to make things easier, using "George," the autopilot, seemed desirable — and indeed it performed flawlessly.

Combined, the calm air, the black night, the use of the autopilot, the aircraft instruments in seemingly fixed position, and Simpson's relatively fixed seat position, eliminated for Simpson any feeling of motion. The crew could have been in a dark room somewhere looking at fictitious, painted instrument dials. The whole scene took on a surrealistic appearance, all the more so as boredom and fatigue started to set in. Simpson wished he could drink some coffee, something that usually kept him alert for an hour or two; but unfortunately and unaccountably, no thermos of coffee had been brought aboard. They all did have, however, box lunches containing sandwiches, fruit, a hard boiled egg, and juice. Simpson hoped very much that they would soon have some kind of contact with a submarine. Such a contact would have the immediate effect of bringing everyone, himself the more so, to life. It was not so much that Simpson had counted on working with submarine contacts to keep himself alert; it was just that he could not believe that there would be none till this point. He felt frustrated and wondered about the crew. He asked his plane captain, Hal Weiss, to walk and crawl back to each crew position to try to get some idea of how they were holding up. Hal reported back that everyone seemed to be in good shape, even if disappointed with the dull state of affairs. Simpson wished he could leave his pilot's seat and check each crew member himself; he felt that, not only would it give him a good firsthand impression of each person's status, but it would also help break the monotony of his position and make him more wakeful. Eventually, Simpson would have to visit each crew position just to stay awake.

By 0330, two hours on station, Simpson was utterly miserable with drowsiness and tried to devise little things to keep alert. He devised a new search plan with shorter times on each heading: They would fly twenty minute legs at right angle to the expected course of the subs, counting each minute as it passed in an effort to remain awake. Worried about the crew, he asked his TACCO, LCDR Gordon, to make more frequent reports; but the reports failed to help much. Now, eight hours after he had had his light pre-briefing dinner, he thought about his box lunch. Would that make things worse as his body digested the food? Or would the act of eating itself, somehow save him from his drowsiness? Sure enough, the food did, at first, provide an interlude; but after a half hour or so, things became worse. In the SP2H Neptune, there just are not any suitable places where a pilot can get up out of his seat and stretch out flat — and still be close enough to the cockpit in case of emergency. But there is a space fairly close: The small passageway below the flight deck and to the left of the nosewheel well, called the "nosewheel tunnel," which leads to the Plexiglas nose cone and MAD seat. There the pilot can lie down on the cold aluminum alloy deck and stretch out fully. It is not a comfortable spot; but under certain conditions it feels wonderful. In the past and on less important missions, Simpson had stretched out there after getting the crew chief to take his seat. How he wished he could stretch out there now — he was certain just ten minutes would be enough to revive him. On long night flights when he had been tired before, he would lie there for ten minutes to refresh himself for the landing. Probably never a good idea, he always felt that the demands of landing a big aircraft and crew in the dark of the hours before sunrise justified his rest. He would never sleep, of course; but the hard metal deck provided a feeling of luxury and an unbelievable, if temporary, restoration. Though Simpson was certain that such a rest now would do the trick, he was concerned about the impression and effect his leaving his position would have on his crew at their positions. Would it convey a lack of interest in the mission? Certainly, he did not want to do that. So he pushed the idea out of his mind. He did, however, resolve to get out of his seat and visit each crew position if there was no action and if conditions did not change. Thinking of his copilot, he suggested to LT Carlson that he do the same.

Now, at 0400 hours, Simpson got out of his seat for the first time to visit the various crew stations. He felt that physically getting up and talking to each crew member would somehow revive him. To visit all the stations, he would have to crawl over part of the aircraft's wing which extended right through the fuselage. The space was not quite big enough to crawl on your hands and knees, so it was necessary to part creep and part crawl, but mostly to sort of slide on your stomach. Simpson felt that the effort would help to wake him up. His getting up to visit did help a bit and he returned to his seat in the cockpit — but the revival seemed to last only a few minutes.

At 0430, with no coffee, no sensation of flight, and no contact with the subs, Simpson thought that his next best hope was for daylight to arrive — and hopefully sunshine. That would give him, and everyone else new vitality. And it should not be too long to wait, even in his agonized state, because the first gray of daylight should arrive between 0530 and 0600. Shortly after 0545, some gray appeared, and with it a ray of hope that drowsiness would depart. Like other miscalculations Simpson had made, he found that the gray twilight was darker and more long-lasting than he thought it would be. Instead of darkness, grayness had taken over; and everything else remained the same: the boringly calm air, the muffled, drowse-inducing engine noises, etcetera. Since the sky was overcast, the ocean and the sky were the same color; and there was no horizon to see, requiring that they continue to fly the aircraft on instruments, perpetuating the feeling that they were not moving but holed up in some kind of simulator on the ground. They still had no coffee, no turbulence, no contacts and no sensation of flight — just a prolongation of the pain of fighting drowsiness. Simpson wondered how others on longer flights had been able to stay awake and alert, knowing full well that another flight under similar circumstances could be completely different.

What was there left to do? Would only full daylight on this gray day bring relief? But that would be at 0730; and based on Simpson's earlier experiences this night, he was not sure of relief then. It was not actually until 0830, when they were relieved on station, that he was fully awake. Eight and a half hours in the air, and they had accomplished nothing — except fight drowsiness. Although it was not for lack of effort, the frustration remained. Now, all they could do was return to base and land at about 1000. As they made their approach to land, everyone was wide awake — not because of landing at home base, but because of the real change in conditions and the perception of real motion as opposed to what seemed like simulated conditions on station. With everyone's being fully awake on landing, it was difficult to believe that Simpson, and no doubt others, had been so miserable and had accomplished so little. There was nothing that they could have done about the hours of the flight or the blackness of the night or the stillness of the air or the absence of submarines in the operating area. Still Simpson thought he could have been better prepared somehow and was determined that the next time would be different.

After landing, and for some time later, Simpson thought about this flight; and his thoughts troubled him. They might as well have stayed on the ground in an electronically simulated flight exercise. But although canned programs were available for smaller aircraft with smaller crews, none existed for such a plane and crew — or for such a long duration. Simpson rationalized that, despite fighting so unsuccessfully the extreme drowsiness, such a flight served a real purpose. The conditions they had encountered were real and could occur again, so it really was a test of endurance and perseverance. Many pilots and crews flying somewhat similar missions out of places like Maine, Jacksonville, Iceland, Bermuda, Puerto Rico, Spain, and many other places must have had similar experiences. So, it really was a learning experience — certainly one to profit from. Placed in a pilot and crew's overall flight experience, it had to be considered worthwhile, because Simpson felt that he had learned a lot: Despite all his early planning, he had not foreseen many things. He would be better prepared next time — perhaps not completely, but better! On reflection later, he wondered with renewed admiration how Charles Lindbergh, flying alone, had stayed awake for twenty-seven hours on his flight from Roosevelt Field on Long Island, New York, to Paris.

Ice Floes and Helicopters — A Rescue Mission

It was a cold, gray July morning in the Arctic, not unlike many in summertime as the Ice Cutter *USCG Eastwind*, located now at sixty-seven degrees and thirty-five minutes North Latitude and sixty-three degrees and thirty minutes West Longitude, tediously made its way through the rotting ice fields well up in the Davis Strait between Canada's Baffin Island and Greenland, heading north into more densely packed ice. Its mission, named "SUNEC II," was to build several — five to six — "Seriesor" unmanned radar stations atop various mountains on Baffin Island, where their antennae would face the Soviet Union across the frozen North. Once in operation, they would scan the skies for incoming ICBM's (intercontinental ballistic missiles) or enemy aircraft formations. Naturally, for the installation of this equipment, it would be necessary to keep the sea lanes open so that ships could supply equipment for construction of the station and perform any required maintenance on other sites in the area. The sites would become part of the many radar posts in what would be named the Distant Early Warning Radar Stations (Dew Line), a chain of radar installations, so important in the dangerous phases of the Cold War, which would extend for three thousand miles eastward across Alaska from Cape Lisbourne in Northwest Alaska, to Baffin Island, Canada, in the east. This, in turn, would be part of a larger system spanning the Arctic eastward to England, all of which would face the Soviet Union across the top of the world. Information on any incoming missiles would then alert NORAD (North America Radar Defense) for action.

The *Eastwind's* position placed it just north of the Arctic Circle (66 degrees, 32 minutes North Latitude) in the Davis Strait, or 3,992 nautical miles (4,580 statute miles) north of the equator, yet about 1500 nautical miles south of the North Pole. The date, July 21, was part of a window of time when it was possible to get and keep the shipping lanes open before the cold weather would freeze them over again for the rest of the year. The choice of this time of year was chosen as one of the best times, the time when the sun is still near its highest point in the northern hemisphere and its direct rays shine down near the latitude of twenty-three degrees thirty minutes north, the The Tropic of Cancer. But if the sun's rays are directly down at The Tropic of Cancer on the twenty-first of June, the rays are slanted, and less warm, towards all other areas. So theoretically, a month before and a month after the first day of summer — that is, from May twenty-first through July twenty-first would be expected to be the best times for warm weather in the Arctic. During this time, the sun in the high north latitudes, that is above The Arctic Circle, remains visible late in the night, sometimes remaining visible all night — "The Land (or ocean) of the Midnight Sun." It is precisely these relationships of the sun's rays that have resulted in naming both The Tropic of Cancer and The Arctic Circle, bringing to mind terms which most people encounter in their early school years and then rarely think about.

But, of course, there would be no seasons or need for designations such as Tropic of Cancer, Arctic Circle, etcetera, were it not for the earth's tilt on its axis — twenty-three and a half degrees — as it made its yearly journey around the sun.

Also, though unnecessary, it might be worth a reminder that the nautical mile, the measure used in both ship and aircraft navigation, is based on the north-south distances along any meridian and corresponds to a minute of arc — or, since there are sixty minutes in every degree and ninety degrees from the equator to the pole, there are fifty-four hundred nautical miles from the equator to the pole. Each minute of arc (a nautical mile) equals six thousand eighty feet. Since a statute mile is only five thousand two hundred eighty feet, each nautical mile equals about 1.1515 statute miles. While we usually think of the earth's circumference as being about 25,000 miles, the 5,400 nautical miles in each quarter sector equal 21,600 nautical miles; and multiplying 21,600 by 1.1515 equals 24,873 statute miles, or about the same measurement.

How far north any ship could travel even in the best seasons and how far north a ship could travel before it ran into solid, impenetrable ice would have to be left to the ship navigation experts. But with the Arctic Circle's indicating the line north of which the sun does not rise above the horizon in winter and which receives only slanting rays of the sun in summer, even if daylight then lasts past ten or eleven o'clock at night, it takes but little imagination to realize that the area north of the circle remains icebound so much of the year. Be that as it may, this was the second year of the effort to build the radar stations; and the particular goal this year was to sail into uncharted fjords and deploy a Navy Underwater Demolition Team (UDT) to clear a beach safe of all hazards for the landing of heavy amphibious cargo boats. In the absence of docks, the giant amphibians would off-load large D-8 Caterpillar tractors, construction equipment, and food and supplies necessary for the work team to build a road to the top of a designated mountain and there install the radar and protective dome. The "frog men" of the UDT would blow large boulders out of the way using M-2 explosive charges.

It was slow going for the *Eastwind* to move toward its desired position in this second year of "SUNEC II;" and the ship needed "eyes" to guide it to any open stretches of water in its path. The "eyes" would be a helicopter, the HTL UR-21.

Beautiful clear weather eliminates many of the complications of flight; but on this July day, even the cold, gray weather did not seem to present any problems. With weather being such an important consideration in all flying, including the military, the weather conditions, though not ideal, augured well for today's helicopter mission off the stern of the ship. Indeed, it was the kind of day when any thought or even a hint that something might go wrong seemed extremely remote. Any military pilot with extensive flying experience knows that things can go wrong at any time and in the best of flight conditions; but good weather can preclude the risks of many types of problems and reduce the number of complications of more urgent problems. So, there was no hint that today's flight would or should be anything more than "a piece of cake." The purpose of this morning's flight was an authorized ice reconnaissance mission; and the aircraft, the Helicopter HTL — UR21, was outfitted with rubber pontoons rather than the usual wheels, or in some cases skids, allowing for landing on the aft wooden flight deck of the ship, solid ground, or ice, or the sea. The pilot would be LT Francis (Frank) Lagan; and aboard the helicopter as an official ice reconnaissance observer was the *Eastwind's* Executive Officer Commander Roger M. Dudley. LT Lagan was an experienced pilot, one who had learned the principles and skills of flying in the Navy's flight program and had them reinforced as a flight instructor at Pensacola Naval Air Station, Florida. Later, going through helicopter flight training, he added rotary wing aircraft qualifications to his previously obtained fixed wing qualifications.

Although starting aircraft engines in cold weather sometimes requires preheating the engine by directing warm air over the cylinders to loosen any congealed oil, that was not necessary; and LT Lagan soon had the engine started and warming up. Engine oil temperature's being one of the critical items in a warm-up to insure that all moving engine parts receive the proper lubrication, it was necessary to be patient while that temperature rose. After a good warm up and checking the engine magnetos to insure there was sufficient spark to provide proper ignition for the fuel-air mixture in all the cylinders, Lagan and the ship's executive officer were ready for takeoff. The preflight inspection, the engine's start-up and magneto checks having gone as he would wish, Lagan's takeoff was normal and they proceeded smoothly on course ahead of the *Eastwind*, climbing and maintaining an altitude of five hundred feet for a good look at the ice and channels below.

With everything proceeding as expected, what could go wrong?. Engine reliability, usually excellent, would seem to portend well for such a routine flight. But Naval Aviators, and other pilots with extensive experience know that engine malfunctions, limited as they are, are always a possibility. So it was on this day, when a mile ahead of the ship and after only about ten minutes of flight, trouble started when the engine quit with a loud BANG! and the complete loss of power, leaving Frank the severe prospect of how much could be done with only five hundred feet of altitude and the forbidding prospects below.

Engine failure in helicopters, provided no engine parts fly off to damage the tail rotor or main rotor, does not preclude dropping down to a safe landing. The best option is to have the main rotor auto-rotate; that is, to let the wind across the rotors caused by the descending motion of the helicopter through the air keep the rotors turning and still provide enough lift so that the descent is slow enough to effect a safe landing. But this option requires a prompt decision and a certain amount of altitude to start with; and while the 500 foot altitude should be sufficient to obtain auto-rotation, that depended on the rotor blades' disengaging from a frozen engine. Faced with this uncertainty and the need to land either in a clear stretch of water or on a sufficiently large ice floe, LT Lagan provides the best description: "It did make a BIG BANG when it quit at about 500 feet. Not knowing what that was, I, as well as any Navy-trained helicopter pilot, went immediately and unthinkingly into a full auto-rotation — with an unspoken prayer! Full down on the 'collective stick' and turn into the wind, all the while looking for a place to set her down nicely." He hoped to control his rate of descent to provide a safe landing wherever he could, provided it was a sufficiently large and level place for that landing, whether that be in the water or on an ice floe. His choice was not without its risks. With no power and not knowing whether something else could occur at the wrong moment, he ran the risk of not being able to land on a level spot. The wrong spot, the very edge of a uneven ice floe, could flip the helicopter over and throw the executive officer and LT Lagan into the 28 degree water — perhaps inverted. That would be worse than landing in a sufficiently clear area of ocean. Although they both were wearing exposure suits and Mae West life preservers, this equipment was quite bulky and would complicate their exiting from a sinking, perhaps inverted aircraft.

The situation Lagan found himself in trying to find a satisfactory landing spot required a very fast, almost instantaneous decision. What seemed best in the second or two he had to react? A soft, safe landing anywhere was a possibility. But what about the risk of tipping over in the frigid water. Even if they could escape from a helicopter tipped on its side, would they find themselves coming up under an ice floe with no chance of survival? How much LT Lagan thought of all his limited possibilities, perhaps even he did not know. He chose to take a chance and try to effect both a landing that would not result in injury and one that would at least keep them dry. It was not until LT Lagan's helicopter had descended to about 200 feet that an ice floe beneath him looked as if it were both big enough and level enough to land on. But it was still too difficult to know with assurance that the floe was level enough. The floe was long enough at about 150 feet, but it was only thirty feet wide, although about eighty percent of it appeared smooth. It was certainly wide enough provided it was smooth and even across its narrow dimension — and assuming he could bring the aircraft's forward movement to a stop in time. The approach he would try to make would be dictated by hoping to land as much into the wind as possible, light as it was at about six miles per hour, and by other things that he saw and considered possible under the circumstances. His decision on landing on an angle across the ice floe would give him a runway length of about seventy-five feet.

Committed now that he was forced to and had made his decision, he aimed toward what he thought was a relatively even area and decided that he would reduce his rate of descent by flaring the helicopter at about fifty feet above the floe. The flare, if executed properly, would reduce both forward velocity and rate of descent and would let him descend the remaining distance at a safe rate. In addition, it would let him know about where he would touch down. Although he could not be certain, he did feel that, since the ice floe was not a sheet of slick ice but rather ice crystals and snow, its surface would allow him to stop before sliding over the edge of the floe — with the uncertain results that would precipitate. Accordingly and at about fifty feet and descending, LT Lagan increased the lift angle of the rotor blades, flaring the aircraft by slowing the rate of descent and forward speed and establishing a safe rate to touch down safely. Although he touched down approximately where he wished, he did have some forward speed. To force and keep the helicopter on the ice, hoping thereby to negate the forward speed, he decreased the lift angle of the still rotating rotor blades to a negative lift position, forcing the pontoons down against the ice. The result was all he could have hoped for, the helicopter coming to a stop quickly and safely, if a bit too close to the downwind edge of the ice floe. They were down safely, the big concern in all aircraft incidents. Both LT Lagan and the executive officer were unharmed; and as they stepped out onto solid ice, they felt lucky, Frank expressing a silent private "Thank you Lord." All they had to do was wait to be picked up somehow. They could not help but recall that on a similar flight mission last year, in the same type of aircraft but with different mechanical problems, both the pilot and the executive officer were killed when the helicopter crashed and fell inverted into the frigid water. Safe and unharmed, they were aware that the aircraft was also unharmed at this point, even if it had a severe engine problem that would require lots of work — provided, that is, it would be possible to repair it and, most importantly, provided it could be salvaged without further damage.

On the ice floe, Frank reflected on the many times he had practiced emergency auto-rotations on routine flights; three times on every flight was standard operating procedure. Such practice provided the instantaneous response he needed on this flight. But what now? Their first thoughts having been about the engine-caused emergency, Frank did not think about how the helicopter might be recovered and repaired on the *Eastwind*. But now, with the handling of the first part of the emergency being all that could be hoped for, and with both Frank and the "exec" more relaxed, thoughts began creeping in, considerations that they had not had time to think about till now. They were now a couple of miles from the ship. Would it be possible to get the helicopter back aboard the ship somehow? How close could the ship approach to them? If the helicopter was lost, what would that mean to the ship's mission? Yes, there was another helicopter on board. But if something happened to that one also, the ship's mission would be seriously restricted. Ordinarily, the loss of an aircraft would not be a big deal. After all, aboard aircraft carriers, severely damaged aircraft, particularly in wartime, are just pushed over the side. But with just two helicopters available, the loss of one of them would be a serious problem. Lagan called the *Eastwind* to report that no one had been injured and that, other than the engine, everything was intact. Similar thoughts of rescuing the helicopter went through both Lagan's and the Executive Officer Dudley's minds, while, at the same time, personnel on the ship too had similar thoughts of recovering the helicopter. After all, if they could get the helicopter back aboard ship, they might be able to make engine repairs; or if necessary, they could even change the engine. At worst, there was the possibility that the helicopter could be scavenged or skeletonized for parts for the remaining helicopter.

But there was a difference of opinion as to how best to get the helicopter back aboard ship. First, the ship would have to get as close to the helicopter as possible. Then some means of getting the aircraft into a position from which it could be hoisted aboard ship would be necessary. However, even if

Where There's a Will 75

the ship could approach reasonably close to the helicopter's position, the boom on the hoist would only extend so far over the edge of the ship. The *Eastwind* maneuvered as best it could to close the distance to the helicopter, only to find their best efforts could bring them no closer that about six hundred feet for fear of destabilizing the ice pack. To the *Eastwind's* boatswain's crew, who were approaching with coils of rope, it seemed best to attach a line to the helicopter and winch it back to a spot from which it could be hoisted. To LT Lagan, that solution seemed both too simple and too dangerous. "It seemed that they would manhandle the 'chopper' back. There was no clear path to do that. The ice field was serrated with deep channels of melted ice. The area in which we had to put down consisted of many small pancake-like ice floes; and if the helicopter were to be winched back, circumstances would require that the helicopter be winched off one ice floe and back on the next, and so forth, until hopefully the helicopter would be close enough to be hoisted aboard." Helicopters are very delicate flying machines, so the boatswains method of recovery seemed too risky to LT Lagan, who felt that surely at one point the helicopter's being moved off one ice floe and on to another would result in tipping the aircraft over and either sinking it or damaging it beyond the on-board aircraft maintenance crew's ability to fix it. Even if the helicopter could be safely pulled off the floe it was on and into the water without tipping over, much risk still remained when they would attempt to pull it up onto the next pancake of ice. There just was no hope or possibility that a straight path through the water back to the ship could be found; and with a line attached from the ship to the helicopter, it would be impossible to find such a path through the water to the ship. Any path other than a straight path ran the almost certain possibility that the helicopter would be tipped over or ripped apart. Any thought of the possibility of further damaging the "chopper" was anathema to Frank, who, in this brief incident, had developed an attachment to it. After all, it had responded well in the emergency, and he did not want to see it "hurt."

LT Lagan did not like the boatswain's method, even though, as a last resort, the *Eastwind's* captain could order it. Frank wondered whether perhaps there might just be another way. Was it possible that the helicopter's engine could be restarted and even, if at best, it could only have marginal power, then be flown back toward the ship? Conferring with a young aircraft crew chief from the *Eastwind* whom he respected, Frank and he decided that, if they could possibly restart the engine, they might have enough power to lift a foot or two into the air briefly to "hop scotch" from one ice floe to the next. His method would be quite risky, risky not only to the aircraft but also to himself. In addition, he would then have to convince the ship's skipper that his idea would work. Whatever the skipper decided posed risk, the risk to life being the most important. Frank was very much aware that his idea of playing hopscotch with the helicopter would be like creating the emergency he had just handled all over again — and again — and again. In hop-scotching he could find himself in a very bad position relative to the next ice floe; and what if the engine did not respond sufficiently to his attempts to increase power? What if the engine failed completely on lift-off just at the edge of the ice floe he was attempting to leave? Any such attempt was far from a sure thing. Perhaps his idea was not such a good one. Perhaps they should count their blessings, even if it meant abandoning or pushing the helicopter into the sea.

Even before LT Lagan received a reply from the ship, he decided that he would be justified in making the decision to go ahead with his plan. With the helicopter some pounds lighter without the executive officer on board, Frank might be able to manage it — provided he could get the engine restarted. By this time, the *Eastwind* had maneuvered a bit closer, creating in its maneuvers a pond of open water on its port side between its hull and the nearest ice floe. There were, however, four ice floes to hopscotch over before reaching the pond.

Restarting the engine was no sure thing; but Frank did manage to get it started using battery power on the first attempt, even if it sputtered and was rough-running. While in place, he managed to check the helicopter's lifting ability, calculating that his odds were not bad and his plan worth a chance. Since the ice floe he was on was flat and the next also, he had no obstacles to clear and could stay in flight a foot or two above the ice and water, a height which, if the engine failed completely, would preclude a crash but which would not insure his not landing on the edge of a floe and tipping over. Frank lifted off the ice he had landed on and, flying a foot or two above the ice and water, carefully moved to the first landing, setting down quickly really not knowing whether he could go any farther but to make that determination after touching down. He could see that the second and third ice floes were flat also; and using the same procedure, carefully advanced. But from the third ice floe, he could see that, on the fourth and last ice floe, there was an ice wall some thirty inches high blocking the route to the pond which the ship had created. Again, moving very carefully, Frank advanced to the fourth ice floe, where he could get a closer look at the problem the ice wall would present. Up to this point, there had been just level ice and water to cross over. The ice wall would require a flight higher, and LT Lagan was not sure his sputtering engine was up to the task. And even if he could clear the wall, could he safely splash the helicopter down into the water and then stop before running into the ship? "I would have to rev her up (the faltering engine), try to jump high enough to clear the top of the ice wall, and then splash down in the water with all the back stick I could give it, hoping that my forward momentum would stop before the side of the *Eastwind* stopped it for me."

Thinking about what was required, Frank went through all the necessary steps mentally. Several times, he rehearsed in his mind what he'd have to do, arriving eventually at the conclusion that he might be able to do it. As planned, he lifted up gradually off the ice, checking to determine whether he might have enough power to lift higher than had been required before. Rather than risk the additional power required over a longer than necessary period of time, he resolved that, if he had the power even for a short period, he would do everything in the shortest possible time. As planned and after determining that he could clear the ice wall, he added the power necessary, cleared the ice wall, and splashed down quickly, hearing the cheers of the ships crew as he hit the water. From there is was an easy matter to hoist the "revered chopper" aboard ship.

The five aviation mechanics aboard ship installed a new engine in a couple of days; and the helicopter finished the cruise without further incident. Everyone aboard ship, including the mission commander, a commodore, was pleased with the satisfactory resolution of the problem. After all, it could have been very serious; and you don't need too many things to go wrong before a mission gets the label of being jinxed. Avoiding the first disaster, as LT Lagan did, went far to avoid that label and helped to bring about a successful mission overall. However, no one knew the risks involved better than Frank, who had to be pleased that his calculations and determination paid off. He was honored by being made a member of the "Old Pro Club," and was written up in the *Naval Aviation News* magazine, though there were some who felt that some further commendation in addition was more than warranted.

Part Three: Human Factors

Human Factors — Stop the Damn Nonsense

Sometimes, situations arise, which though handled well eventually by the pilots involved, raise the question, in hindsight, about the original undertaking or about how conditions, either by choice or happenstance, deteriorated to the point that they became problems. It is easy, for those not involved and who review the situation from the comfort of their living room reclining chairs to pass judgment on others. That is not to say that the reviewers are unforgiving when they say that the situation was pure nonsense and should never have occurred. One need only remember how many long months it takes for the National Transportation Safety Board (NTSB) to determine what happened in a situation in which the pilot had only a few seconds to make a decision. But most aviators will admit that there were a few times in their careers when they wished that they had never gotten involved in some situations and wonder what options they might have had to avoid them. No doubt, psychology played a part, a part which manifested itself differently according to the pilot's background, experience, or character. Nevertheless, no one could really support conclusions that any of the problems which evolved and are described in this section were the result of good judgment. Though the incidents are quite different, they all have a common thread in that personal factors involving pride — or the lack thereof; a sense of responsibility — or the lack thereof; courage — or the lack thereof; and the degree of skill — or the lack thereof, play an important part.

In the first instance, overconfidence and cockpit communications nearly caused a tragedy, even if the pilots were capable and their skills helped avoid very serious consequences — but they were lucky too! In the second instance, the pilot knew he was disregarding good judgment and common sense; but his sense of pride got in the way, resulting in his yielding to the wishes and pressure of a passenger. In the third instance, a pilot's apprehensions or lack of character resulted in his having other pilots carry his load while the air group was operating from the aircraft carrier. And the fourth incident, while somewhat humorous and not really dangerous, resulted in a rough, noisy couple of hours and the loss of a lot of equipment.

Perhaps underlying all of these incidents, is a desire on the part of pilots to accomplish something — to have something to show for their efforts. If you also add in external factors and responsibilities not related to flying, mistakes due to lack of concentration can creep in.

Why Do They Do It? They Know Better!

Of course they knew better. Why would two good, intelligent, experienced pilots take the chance. They knew all about the risks involved in what they were doing. Everyone who flew the SP2H Neptune knew that the course of action these pilots were about to take was not advisable. It was not that fate conspired against them. On the contrary, many things worked to their advantage — with no credit to them.

The SP2E (formerly P2V-5 and later to be simply the P-2) Neptune was a rare bird because it had two Radial R-3350-30WA or 32WA reciprocating engines manufactured by the Wright Aviation Corporation and two J34-WE-34 turbojet engines manufactured by Westinghouse. Taken all together, these engines could produce about twelve thousand horsepower; the reciprocating engines produced three thousand horsepower apiece, and the jet engines, under certain conditions of speed and air pressure could produce three thousand horsepower too. The common expression when using all the engines was "Two turning and two burning." Compared to many automobiles with their one hundred to one hundred fifty horsepower engines, this power seems very impressive — and it is! But compared to the power of jumbo jet aircraft with their forty-five thousand pounds of thrust (roughly 45,000 horsepower) per engine, it seems much less impressive. Nevertheless, it was a lot of power for this particular aircraft. The SP2H airplane was a marvelously stable and versatile airplane with excellent handling characteristics. Ungainly looking and unattractive as it sits parked on the ground, particularly with its bomb bay doors open and hanging down, in flight it is a beautiful bird viewed from any angle (See photo). But despite these fine descriptions and considering certain configurations and characteristics, it, like all military aircraft, had to be treated with respect.

A much earlier, lighter two-engine version of the aircraft became quite famous for holding the non-stop, non-refueling distance record. The airplane involved, "The Truculent Turtle," was flown from Perth, on the west coast of Australia to Columbus, Ohio, a distance of about 11,235 miles. This early, relatively simple model of the Neptune became heavier and heavier as one black box of electronics equipment after another was added to improve its mission capabilities. Before too long, it became too heavy to fly on two engines and still provide the satisfactory single-engine performance required should one engine fail or have to be shut down. The solution was to add two three thousand pound jet thrust engines, one on each wing farther out than each reciprocating engine. The airplane, now configured with its two reciprocating engines and its two jet engines, had plenty of power — and power to spare in emergencies. It became the Navy's only operational — that is, used in operational squadrons — aircraft so configured.

This combination enhanced the performance and operation of the SP2E in many ways. But it was this configuration of both reciprocating and jet engines that presented the pilots with choices so tempting that they caused problems on many occasions. The newly-added jet engines were to be used only for takeoffs, landings, and emergencies such as the failure or shut down of a reciprocating engine. Used as intended, the performance was wonderful: on takeoff, the power was truly impressive; in emergencies, the jets could be quickly started and would provide more than adequate power should one of the reciprocating engines malfunction. With such use, there would be little likelihood of developing fuel problems. With such intended uses, the jets obviously were not to be used in the primary mission of long range patrol and antisubmarine operations. For mission flights, the crew would consist of twelve members: the pilot, copilot, plane captain, tactical coordinator (Tacco), radio man, ordnance man, and others necessary to operate the antisubmarine detecting equipment. For non-mission flights a minimum crew of four was required: pilot, copilot, plane captain, and radio man.

With but little thought, one can see that such an aircraft cannot use both types of engines at the same time under all circumstances. Reciprocating engines cannot be used effectively at very high altitudes, although by using the engine superchargers, they can fly quite high, as the B-17's creating their high altitude contrails in flights over Europe in World War II readily demonstrated. Jet aircraft are designed for high altitudes and are most efficient fuel-wise at those altitudes. So, basically there are two different types of engines on the Neptune, neither operating efficiently at the altitude designed for the other. Yet each was complementary to the other in the uses for which intended. Although the reciprocating engines had superchargers, the Neptune rarely operated at altitudes requiring their "high blower" use. However, a more practical device was a system of turbines called "blow down turbines" to recover power from the engine's exhaust. These turbines were built into the engines and were in operation at all times, providing lower fuel consumption "for the same power of the basic engine or higher power for the same fuel consumption." Automobile manufacturers have, from time to time, incorporated a modification of the "blow down turbine" on some engines and models, usually using the term "turbo charger," apparently a somewhat misleading appellation to convey a sense of power.

As great as the addition of the jets was, the addition came with a price. The airplane became much heavier, requiring more fuel and the addition of wing tip fuel tanks to hold the additional fuel. The airplane also became more complicated. Nevertheless, when used properly, it was a terrific airplane overall. Problems with fuel arose when pilots would use the jet engines at other than their intended times. This misuse would usually occur when the pilots would keep the jets on after takeoff, trying to build up and maintain higher than their normal cruising speeds to get from one place to another more quickly. Such use was not authorized and caused more problems than it solved. Although there have been many instances of misuse, the experience of one crew provides more convincing proof of misuse of the jet engines than a mere mention of the misuse possibly can.

CDR Bill Lewis, Copilot LCDR Harry Wren, and their crew had hoped to get airborne at 1900 hours from the Naval Air Station, Olathe, Kansas, en route to their home base at the Naval Air Station Willow Grove, Pennsylvania, in their SP2E Neptune aircraft. It was important to get airborne by 1800 hours Central Standard Time for several reasons, the most important being that the air station at Willow Grove would close down at 2400 Midnight Eastern Standard Time, or 2300 hours Central Standard Time; and it was unrealistic to think that they could make their destination in much under five hours. The first problem arose when a delay occurred in refueling the Neptune, which they had brought in from near the West Coast a couple of hours earlier at about 1600 hours. The delay in refueling at Olathe meant that they could not realistically get airborne out of Olathe before 1900 hours Central Time — 2000 hours Willow Grove time — and, as a result, could not hope to arrive at Willow Grove before the field closed. The approximately nine hundred nautical miles (1036 statute miles) would require very close to five hours of flight, since they could not expect to make more than two hundred knots (230 miles per hour). Thinking about what to do in their circumstance, they decided that they would file their instrument flight plan for four and a half hours en route, get airborne as soon as possible, and keep the jets on for awhile to come as close as possible to the scheduled time en route. It was a decision based on what seemed reasonable, but which would require that a close observation of their fuel supply be maintained. Underlying their decision, were a couple of assumptions the pilots hoped would prove satisfactory: the weather was expected to be good all the way; westerly winds could usually be counted upon to improve the ground speed; and Willow Grove would probably stay open a bit after midnight if they had received a report on the flight's expected arrival time shortly after Lewis and Wren became airborne — surely, they would remain open for just one half hour beyond their normal closing time for one of their own aircraft. When making assumptions, it's hoped that they will all prove to be correct. Even if most of the assumptions work out, we can hardly expect that we'll always be right and to the degree hoped for; that means that where several have been made, some might be quite good — but then some would be wrong.

With Wren now in the left or command pilot's seat, since Lewis had flown on the flight from near the West Coast, takeoff was at 1855 hours Central Time, a bare four hours and five minutes before Willow Grove would close for the night. While the takeoff time was not what they had hoped, it was still reasonable to assume that their plan would work. The climb to their assigned altitude and their leveling off at their cruising speed were normal — except that at no time after takeoff did they shut down their jet engines. The result was that their cruising speed was about twenty knots faster than with using the reciprocating engines alone. Although they were about as high as they could be (7000 thousand feet) without using oxygen, which they did not have, the altitude was far below the altitude for efficient jet engine operation including fuel economy. The big decision Lewis and Wren had to make was to determine how long to keep the jet engines on, yet still have enough fuel to make a safe landing, perhaps even to get to an alternate field should the weather suddenly deteriorate or the field be closed, and allow for unforeseen problems. If they could maintain two hundred twenty knots, or more with a favorable wind, they could make the trip in just about four hours. Lewis and Wren were good, experienced pilots, with good cockpit communication skills — they provided good cross-checks on each other. They knew that they could not maintain the speed they were making; and as the flight progressed, they saw just how much their cruising speed would be reduced by minor, but important, developments. Favorable winds at their flight level were very light or nonexistent, when a good, strong tail wind would have virtually assured their arriving nearly on time. In addition, never having kept the jets on for trips before, they had no realistic idea of just how fast they were using up their fuel. Looking at their fuel gauges, along with Plane Captain White, they felt that, above everything else, they did not want to get into one of those situations they had heard about. Therefore, approaching Columbus, Ohio, they decided to shut down their jet engines so that, even if late, they would insure a safe flight. After all, they could always land elsewhere if necessary.

Approaching Pittsburgh, they realized that they had not been prudent enough. The fuel gauges now seemed to indicate much less fuel than when they shut the jets down approaching Columbus. Perhaps they had misread or miscalculated. Passing Pittsburgh, they realized that they would not have as much fuel upon reaching Willow Grove as they had planned for prudent, safe operation. Despite what they had thought were careful and reasonable considerations, they had indeed kept the jets on too long. But even in their situation, there were some encouraging indications: the forecast weather was holding up, relieving them of the necessity of making a time-consuming instrument letdown. The beautiful clear skies with their bright stars were uplifting. But crossing Harrisburg, it became very clear that the fuel situation would become serious. Nevertheless, rather than dropping down and landing at Olmstead Air Force Base in Harrisburg, they continued on and soon were receiving strong signals from Willow Grove's TACAN navigational system and, before long, the TACAN indicated that, as they crossed Lancaster, they were fifty nautical miles from Willow Grove. Fuel, it seemed would probably not cause an emergency problem, even if they had to be very careful. One concern, of course, was whether all of the relatively few remaining gallons would be usable.

Although Lewis and Wren were very experienced, they could hardly have expected a couple of things which now were to cause even more serious problems and concern for potentially disastrous results. Crossing Lancaster, Lewis called Navy Willow Grove for landing instructions, at the same time cancelling their instrument flight plan in view of the good weather and in the interest of saving time — that is, fuel. Now, very close to Willow Grove, having received landing instructions, and still receiving strong TACAN signals, all they had to do was simply to follow the needle on the TACAN to home in on the base. But unknown to Lewis and for some unknown reason, Wren, not seeing Naval Air Station Willow Grove in the profusion of lights below, passed right over the field without seeing it and without telling Lewis that he did not see the field and did not know where it was. Calmly continuing straight ahead for about a minute, Wren started to make a wide circle while he looked for the air field, still not saying anything to Lewis about his not seeing the field. The wide circle and the absence of anything from Wren made Lewis think that Wren, under the circumstances, was being too casual in getting ready for his landing approach. When Harry Wren continued to move away from the field, Lewis became extremely uneasy considering their low fuel state. When he could contain himself no longer, Lewis, not wishing to alarm the crew by using the intercom, yet revealing his own sense of worry, leaned over close to Wren's helmet and said, "What in the hell are you doing?" The answer was a classic example of a problem, more common than usually thought, of poor communications between pilots in dual-piloted aircraft. Wren's response to Lewis' question was, "I lost sight of the field." It was only at that point that Lewis had any idea of what Wren was doing. There were many bright lights on the ground in the vicinity of the field this clear night; and though unusual, his losing sight of the field was somewhat understandable. But what was not understandable was his not saying anything about it. It would have been such an easy thing to ask Lewis if he had the field in sight — and much more in keeping with his nature and the good rapport that existed between the pilots. But for some unknown reason, he did not ask; and the precious time lost became a factor in what still lay ahead.

Having seen the air station about fifteen miles before they passed over it, Lewis was able to immediately point it out. Wren turned in an instant and then turned downwind parallel to Runway One-five (150 degrees) following prescribed procedures. Precious fuel and hence the minutes of flying time that that fuel would have provided had been lost. Wren turned on the jets as he approached the turning point at the downwind end of the runway. Although the fuel-saving desirability of not turning the jets on had been considered, they felt that they had better follow recommended procedures — and besides, they were so close to landing that they felt they could turn them on. As soon as the jets were ignited, Wren placed the wheel lever in the "down" position and adjusted his wing flaps to the desired position, thinking that, at this point, they were less than two minutes from touchdown on the runway and a very instructive flight would come to an end. But wait! A routine check of the wheel indicators by both pilots revealed that, although the main wheels were down and locked into position, the nose wheel was not. LCDR Wren continued his approach hoping that, by the time the aircraft came out of the approach turn and was lined up into the straight-in final approach for landing, the nose wheel indicator would show "down and locked." No such luck!

When the nosewheel indicator still did not show "down and locked," they had no choice but to take a wave-off and make another approach. Yes, they could have landed with the nosewheel unlocked and sustain some non-life-threatening damage to the aircraft. But they felt that they still had a little fuel and preferred to think of a landing with an unlocked nosewheel as a last resort. However, with the fuel at a critically low state and not knowing how much time would be required to get the nosewheel "down and locked," they knew their situation was getting "iffy." The jets were now on, the aircraft was in a high drag configuration with the main wheels and the flaps down, and they had to add fuel-consuming power to get up to the required downwind altitude. To reduce drag and save fuel, they raised the landing gear and did not advance the jets above the "idle" RPM position. The two reciprocating engines easily providing the necessary power, Wren climbed to the proper altitude, turned downwind for another approach, and placed the landing gear lever in the "down" position, hoping that the raising and lowering — or recycling — the landing gear would somehow result in getting all three wheels "down and locked." Both Wren, Lewis, and the plane captain knew that they had a hydraulic leak somewhere that was causing the problem. They also knew that, even if the nosewheel did not come down, there was a manual solution which would get the gear down. The problem was that they did not know if they had enough time to use the manual solution, and so they opted to just try once again to get the gear down hydraulically. After Wren put the landing gear handle in the "down" position, the main wheels were soon in the "locked" position; and perhaps somewhat to their surprise, the nosewheel popped into the locked" position a few seconds later, being verified by the nosewheel indicator. Harry Wren started his approach turn to Runway One-five again, concentrating on flying the aircraft, at the same time hoping there was enough usable fuel to complete the landing approach to a safe landing. He made his final adjustment of the flaps, while the plane captain, always watching the fuel gauges, stood by to make some quick lever changes to get the last bit of fuel from another tank that might just have a tad more usable fuel. As they turned into the final approach to the straightaway, CDR Lewis reminded Wren, if he had to be reminded at all, that the hydraulic problem that caused the sluggish nosewheel operation could cause further problems on landing — mainly braking problems. Now over the threshold of the runway, Wren eased the plane onto the runway in a very smooth manner, despite the unsettling events of the last several minutes. Shortly after touchdown on the runway, Wren placed the propellers into "reverse pitch" to brake the speed of the aircraft before he attempted to use his brakes, feeling that he probably had just a few applications of the brakes before he had no toe brakes at all and would be left with only the emergency hand brake. Wren's concentration had been intense — some things really sharpen the mind — and his approach, landing, and use of reverse pitch had been such that only a slight application of the brakes had been necessary. Nevertheless, both Wren, Lewis and Plane Captain White felt they had been quite lucky to have gone through what they had and still have some normal braking left. As they slowed down and turned off the runway and onto the taxi strip, LCDR Harry Wren and CDR Bill Lewis should have counted their blessings, stopped, and shut down their engines. But they continued on, turning off their jet engines and raising their flaps as they progressed to the flight line for parking. Approaching the flight line, Harry Wren was taxiing very slowly in view of the hydraulic situation, reminding himself that he could lose all braking action at any time. Then it happened!! As Wren was taxiing close to the other parked aircraft, he uttered an expression he subconsciously relied on and yelled, O Jesus Christ," at the same time reaching down to pull the emergency brake handle, which he had never before had to use. But he found the handle, yanked it, and the aircraft came to such an abrupt stop that Plane Captain White was thrown out of his seat and against the instrument panel, the aircraft being stopped so suddenly that it rocked fore and aft for a few seconds, as the full power of the emergency brake accumulator came into full play. This flight had come to an end — an end with relief and much to think about. White was not hurt thankfully, nor was there any damage to the aircraft or to other aircraft on the flight line. Despite the relief and the need to review the events, there was no time to reflect now. The hour was late, the aircraft had to be refueled and tied down, and everyone needed to get some sleep because tomorrow would come early, bringing with it a long flight over the water as part of an important antisubmarine warfare exercise. For now, however, Lewis and Wren were embarrassed to have gotten themselves into such a dangerous situation. While some would suggest that there were no positives in this experience, the pilots remembered that their calculations and assumptions about reaching their destination had been valid; they remembered also that neither of them had lost his cool and that, despite their lapses, they had worked well together to get out of a bad situation of their own making. Lewis could not help but admire Wren's concentration and skill in bringing the SP2E down smoothly and his handling the anticipated braking problems under such stressful circumstances. Lewis and Wren were to have many more flights together — achieving admirable results, but never having nearly so much excitement again.

Many people, including pilots, reading about this and other flights with nearly complete fuel exhaustion will be inclined to ask why the pilots let the situation develop to such a point. Indeed, some squadron commanders might say that they would not let anyone in their squadrons get into such close situations. But in many of the incidents in this book, squadron commanders, many of whom went on to positions of great responsibility, were involved. Perhaps it is because these people have confidence and are used to getting things done. Also, keep in mind the point that these incidents involved many different people not the same people in all incidents. Keep in mind also that these pilots for the most part kept their cool and figured out satisfactory resolutions for their problems. In addition, there is no doubt that other pilots, both Navy and Air Force, will think that these situations are tame compared to incidents they know about — after all, sometimes pilots do exhaust their fuel and have to either bail out or make forced landings, often with tragic consequences. Nevertheless, questions about why and how pilots get into these situations are valid; but no one will ask these questions more often than the pilots involved, because the memory of the incident will remain forever and the incident will not ever be repeated by the same pilots.

Was This Trip Necessary?

The command of "BAT — TAL — YUN! — AT — TEN — SHUN!" from the Wing Commander echoed in the huge hangar of the Naval Air Station, Willow Grove, Pennsylvania, and resulted in a sudden and dramatic change. The officers and men from all of the squadrons — fighter, long-range patrol, antisubmarine, helicopter, and supporting units — loosely arranged in familiar "Parade Rest" groupings by squadrons till now, came smartly to attention in impressive, formal military order. The Wing Commander, who had given the first command, followed up with the command "Dress Right — Dress!" to get a uniform spacing between men and then commanded "Front!" Although the men were so arranged that these additional commands were not really necessary, procedure called for them; and now the units had the look of a large military organization ready for inspection. Indeed had this been an inspection, the commands — and the results — would have been the same. In response to the next order, "Squadron Commanders Report," each squadron commander saluted and presented his brief report concerning those in his squadron present and ready for drilling. This done, and the brief comments by the battalion commander completed, the battalion commander ordered, "Squadron commanders, take charge and carry out the Plan of the Day!" This formality officially began the drill weekend for the hundreds of naval reservists in the aviation units. Across the hangar Marine Corps units were going through their own formalities. These procedures, far from being onerous or boring, served an inspirational purpose and gave each man a sense of pride and a wish to do his best.

After the battalion assembly and muster were completed, the officers started for the ready room cheerfully greeting their squadron mates, whom they had not seen since last month's drill; and, quickly forgetting any tiredness from just completing a week's work as lawyers, accountants, educators, salesmen, airline pilots, students, etcetera, they were ready for what lay ahead. The drill weekend, sandwiched between two civilian workweeks, would mean that each person would be working twelve days in a row — and this every month! But the formalities, the greetings, and the subsequent informalities gave them new energy, pushing any thoughts that they might be tired and in need of relaxation out of mind.

The officers had hardly reached the ready room, when someone yelled, "Coffee break," and received the laughs he knew he'd get. His remark seemed all the more humorous, because the pilots knew what kind of weekend was in front of them: two flights on Saturday, including a night flight, and one possibly two flights on Sunday for a total of at least ten hours; and in between, there would be briefings, lectures, flight simulators, and the inevitable and eternal paperwork. Sure, there might be time for a coffee break; and there might even be time late Saturday night for a couple of drinks and a convivial snack or possibly dinner with four or five squadron mates.

Though it would be a long weekend in terms of hours worked, the time would speed by. The pilots would seize each pause, each break with enthusiasm to talk about things just remembered. What was it that made these weekends, long as they were, enjoyable? Was it the men? Surely, the guys were a great bunch from all walks of life. But it was not the areas they represented so much as the fact that they had had a similar background in the Navy, a common, binding experience of flight training, aircraft carrier work, and experience in somewhat similar active duty squadrons. What difference if they had not been in the same units when they were on active duty having their experiences. Perhaps it had been that their training and experience had come closer to calling on all of their potential and resources than anything else they had done. And the drill weekends always rekindled thousands of memories: good pilots alive and dead, training sites like Pensacola and Corpus Christi, pretty girls nearly forgotten now that most of the men had been long since married, crazy sea stories, aircraft carriers such as the *Philippine Sea*, *Saratoga*, *Enterprise*, *Midway*, *Coral Sea*, *Roosevelt*, and many others; there would also be war stories from those who had seen combat duty; and, of course there would be the innumerable and inane comparisons of various types of Navy aircraft.

Gathered in the ready room, the pilots listened to the squadron commander's comments, the flight safety briefing, and the operations officer's explanations about the scheduled flights for the weekend. It was only then that LT Pete Larsen's perception of his flights changed from a vague awareness, registered only dimly in the back of his mind by events so far, to a more sharply defined responsibility — to something which he knew would occupy his attention and energy for a good part of this day. Two flights on Saturday and one on Sunday afternoon, each being a little over three hours duration, with only lectures and the next month's flight schedule to prepare Sunday morning as his chief non-flying duties this weekend. Pete Larsen was scheduled for an instrument training flight Saturday morning in the SNB Beechcraft, an S2F Tracker searchlight flight at night, and an S2F Tracker antisubmarine tactics flight on Sunday afternoon. Pete thought the assignment good because of the variety of activities included. Having had extensive instrument flight experience and having been designated the squadron's instrument flight instructor, Larsen was thought to be an ideal choice for that important position. He was responsible for both taking squadron pilots on instrument training flights and also, as the pilots' instrument ratings expired each year, to take the pilots on "Check Flights;" that is, to monitor the progression of training and later to examine their proficiency in a formal flight check. Larsen enjoyed this responsibility because it gave him the opportunity to remain current in the SNB Beechcraft, a good, small utility airplane with frequent challenges. It required some skill to learn to master landing the SNB smoothly because it could easily become a bouncer, exhibiting a rather strange appearance to observers, not to mention the effect on passengers, who were not sure just how the landing would eventually turn out. But Larsen had long since mastered the proper technique and liked to fly the airplane. There were other challenges also; these largely had to do with crosswind landings and rudder control in both crosswind landings and takeoffs. These challenges were well-know by Larsen; and he felt quite comfortable also in these "rudder" situations, even in challenging crosswinds.

Pete Larsen liked this schedule, forgetting for the moment that he had made out this weekend's flight schedule himself during last month's drill weekend. Pete and his copilot, LT Harry Kinsey, who would get the instrument practice, felt they had time for a quick cup of coffee in the canteen where they could discuss what was to be accomplished on the flight. Sipping their coffee, Pete was aware of a good feeling, almost a euphoria. Pete wondered, "Was it the nice clear day indicating the end of the long winter? Was it the effect of the coffee, providing a temporary boost? Never mind the cause. The feeling was there." As they finished their coffee, Pete Larsen glanced out the canteen window to see a very bright March day, a beautiful clear day reducing flying to its simplest. But an expression of brief concern passed over Pete's face as he noticed that the wind sock, which was hanging limp earlier, was now sticking straight out, seemingly rigid, indicating about twenty knots (23 miles per hour) of wind. A quick, enquiring follow-up look at the tetrahedron in the center of the airfield confirmed what was only apparent from the windsock: the strong wind was blowing from the right of Runway Three-Three (330 degrees). Although a crosswind from the right, the direction and force of the wind would present no real takeoff problems, even if the force picked up and the wind varied quite a few degrees from blowing down Runway 33, the takeoff runway.

Proven and reliable, the SNB Beechcraft had performed a variety of jobs over the years; but now, near the end of its career, its use was somewhat restricted, being confined to instrument training and short-haul utility flights. Its tricycle landing gear with a tail wheel marked it as an obsolescent aircraft, most of which had given way to the safer nosewheel design, which have less tendency to "ground loop" in a crosswind landing if the aircraft is drifting slightly left or right across the runway at the moment of touchdown. But it was just this feature and its reputation as a "bouncer" on landing that made this airplane a delight to master. It required considerable skill to "grease it on the runway" without the slightest bounce and then to let the tail wheel down to the runway and guard against any tendency to swerve, a swerve which could get out of control and result in a ground loop. A good stiff crosswind made the task much more difficult. A smooth touchdown in this light airplane, particularly in a brisk crosswind, required more skill than landing the big ones, the jets, or just about any other military aircraft and usually gave the pilot a feeling of accomplishment and satisfaction.

As Pete Larsen and Harry Kinsey slipped into their flight gear, LCDR

Tony Capizzi, the squadron's operations officer, approached them to say that the Air Station had requested that, as part of this instrument training flight, the pilots make a stop at Martin Aircraft Company's airport near Baltimore to drop off a Commander Greville, whom neither Pete nor Harry had ever seen. This modification to their plans seemed to present no real problem since they would be flying on the airways in simulated instrument flight anyway; and they could get an extra landing and still have time for special instrument work such as letdown procedures and basic airwork on their way back.

As they filed their flight plan, Pete could see the repeat wind gauge in the Aerology Department, which now indicated twenty-two knots (just over 25 miles per hour), with gusts up to twenty-seven knots (31 miles per hour); and the direction now slightly different at zero one zero degrees. No problem for the home station with a thirty-five to forty degree crosswind, but what about Martin's company airport? A check of the aeronautical publications showed that Martin Field's main runway was Three-Two (three hundred twenty degrees) or coming from the other direction, One-Four (one hundred forty degrees). Further, there were a total of four runways, making eight directions to land in, depending on the wind; with eight directions, no runway would have more than a twenty-two and one half degree crosswind on it, and, most likely the crosswind would be less than that. A routine check of the *Notices to Airmen* revealed no notices concerning Martin's field, so they set out to fetch Commander Greville and head for their plane. They greeted CDR Greville, noticing his breeziness and easy familiarity and his appreciation for the convenience about to be provided.

With the strong wind steady at twenty-five knots (28.7 miles per hour) and the strong wind, now slightly changed at 015 degrees, the small 8,750 pound Beechcraft traveled what seemed but a very few feet before it leapt into the air with its crew of two pilots and one passenger. As soon as the wheels "clunked" into the "up" position, Greville unfastened his seat belt and moved close to the cockpit so that he could see forward through the cockpit and ahead and become a part of the flight. He exuded confidence as he talked in a very friendly manner; and you could see that, on this bright, clear, bumpy day, he was enjoying every second of this flight as he pointed things out en route, at the same time trying not to disturb Harry Kinsey's concentration as he flew his simulated instrument flight. It may be the pride that some pilots have that make them feel they have to contribute something when others are flying. Most of the time, it takes the form of small-talk, sharing ideas or information about procedures which have proved reliable in certain situations. In Greville's case, he outranked the pilots; and this may have added to his need to contribute — his need to show that he was skilled and competent, knowing that he would not have an opportunity to demonstrate it through actual piloting on this flight. Greville spoke cheerfully about many things, most of which, though ostensibly small-talk, tended to come through as boasting.

Fifteen miles north of Martin's airfield, LT Pete Larsen called Martin Tower for landing instructions. The reply of "Runway Three-two" (320 degrees), the long and most frequently used runway, seemed reasonably consistent; but the rest of the tower's information seemed less so as the tower operator said, "Wind zero-two-five degrees, twenty-seven knots, gusting to thirty. Use caution." Surely there must be some mistake. A sixty-five degree crosswind of twenty-seven knots (31 miles per hour), gusting to thirty. "Why not Runway Zero-Two-Five (025 degrees)?" was Pete's instant thought. There was a Runway Zero-Two-Five, which, though shorter, was more than adequate for the SNB. That would be directly into the wind. Pete called Martin Tower requesting Runway Two-Five, a seemingly logical and reasonable request. But before the tower could answer, the reason for Runway Three-Two became instantly apparent: airplanes from the factory were parked along fully two thirds of Runway Two-Five and Runway Zero-Four-Five, making them unusable — Pete and Harry immediately thought about the publication *Notices to Airmen*, neither recalling that he had seen anything about Runway's Zero-Two-Five being closed. Runway 32-14 was the longer, preferred runway for Martin because of its length, the others being used less often — even if being preferred by pilots flying light aircraft. The situation, which only an hour ago had been such as to reduce flying to its simplest, now presented a real problem: though Larsen could not remember all of the prohibitive combinations of crosswind directions and forces which were illustrated in the pilot's handbook to preclude landings, he knew for certain that the present combination exceeded the maximum allowable crosswind component for the tail-wheeled SNB Beechcraft — the point at which a landing is extremely hazardous and should not be attempted. (Actually, the handbook indicated that, for a steady sixty-five degree crosswind, the maximum allowable wind force was twenty-two knots or twenty-five miles per hour.)

The situation now facing LT Pete Larsen caused many thoughts to race through his head in an instant, all of which seemed to bring him back to the Navy's "Good Sense Rules," which were stressed over and over again and which applied to various dangerous problems pilots confronted. In this case, the rules indicated that it would be unwise to attempt a landing. Of the thoughts going through Pete's mind, one essentially was that, if only this were one of the heavy jet aircraft he had flown or if it were one of the bigger, heavier aircraft, this would be much less of a problem. Pete decided that, with the additional potential hazard of strong gusts, it would be a poor decision to attempt a landing at Martin and resolved to complete their training flight, landing only back at Willow Grove. CDR Greville would just have to make other arrangements.

Greville had been alert, but silent, during the radio transmissions but now came forward confidently to say, "I've been in and out of here many times. Why don't you let me try the landing?" Hearing Greville's remarks, Pete did not answer the question immediately but tried to resolve the problem silently. "Would it be foolhardy to attempt a landing in such conditions. Surely, it would be disregarding strong indications of a hazardous condition and serious trouble. And what about just good plain judgment? And what about Greville? Did his confidence mean that a landing would be simpler than Larsen had thought and could be made without difficulty? Greville was really an unknown until an hour ago — and they still did not know him. Was he as good a pilot as he would have you believe or his self-assurance would indicate? Or was he like some people who appear to be 'checked-out guys' by manner of speaking, self-confidence, and uninhibited personalities — until you get to know them in tight situations and find out that they can't really produce at all. Besides, what if Greville should 'cream' the airplane on the runway? That would be worse than wrecking it yourself because you were the assigned pilot and had signed for the aircraft regardless of what happened." Pete never seriously considered Greville's offer to land the aircraft at Martin. It was completely out of the question. And Pete really would have had no hesitation in telling Greville,

"Sorry but too bad! We're not going to try to land here." After all, dropping Greville off at Martin Airport was just a matter of his convenience — there was no urgent need.

But Greville's remarks affected Pete Larsen enough to rethink the possibility of landing at Martin Airfield despite the dangerous conditions. He wanted a little time to think a bit, but even now not really considering the possibility of attempting a landing at Martin. But as the seconds passed, a plan appeared to form in Pete's mind from things that had been racing through his thoughts just moments ago. Contrary to his best instincts and logic, Pete thought, "Why not try a landing? After all, the wind as imperious as it was, was the only factor to be considered: there was no letdown to be made through bad weather, which would reduce the time for corrections and lining up with the runway when you broke below the clouds; there was no ice or snow on the runway so braking action would be good; the runway was long enough and had no obstacles at the upwind end, so that a wave-off, if necessary, could be accomplished easily; and there was nothing mechanically wrong with the airplane." Nevertheless, it would remain a serious challenge — a challenge, after a change of mind, that Pete now decided to accept.

Very belatedly, Pete responded to Greville's offer to make the landing with, "No thanks. We'll try it one time. If it's too difficult, we'll have to take you back. Sorry." Of course, Pete Larsen never did intend to surrender control of the aircraft to Greville, and he felt that, in making one approach, he could get a real feel for the conditions that existed. Strangely, Greville seemed somewhat relieved. Perhaps it was the thought that an attempt would be made to accommodate him by at least trying one landing. Or perhaps it was because he would not be put on the spot to prove the skills which, till now, he had only boasted of?

Now that the decision had been made, Pete thought about the details of the approach, realizing that a sudden sixty-five degree crosswind gust of thirty knots (35 miles per hour) at the wrong moment could turn everything into worms. There were three main considerations to take into account: first, to touch down safely, there would have to be an adjustment to prevent a sideways drift — a situation similar to that of a driver's sliding partially sideways without control on an ice-covered road, who, when he had slid to the end of the ice patch and hit an ice-free section of road, could turn over and perhaps even roll over a few times if his speed was great enough; second, there was the need to prevent a swerve or weather vaning into the wind when the tail and tail wheel came down to the ground; and third, it was necessary to make a fairly good landing without a bounce, because a bounce which put the plane back into the air would require not only the problem of controlling the bounce but also the drift that was likely to occur during the bounce. Pete knew that he would have to compensate for the strong crosswind from the right, which would blow him sideways to the left. To compensate, he would have to keep his right wing slightly down in a "slip" into the wind to keep from drifting across the runway to the left. He thought about the touch down on the runway and how, at the right time, before the rudders became ineffective, he would bring the locked tail wheel down to the ground in the hazardous crosswind. The tail wheel is being locked would prevent the tail wheel's turning against the tendency of the aircraft, once on the runway, to swivel into the direction of the wind, the first indication of a ground loop. He knew that avoiding a bounce on landing would aid him substantially and that the most difficult part after that would be easing the tail wheel down to the runway and then guarding against a swerve as the rudders lost their effectiveness as the aircraft slowed down in the landing rollout. It was important to keep the aircraft headed straight down the runway both during the straightaway before touch down and once on the ground. (See "More about Crosswinds" at end of this section.)

Pete reasoned that he would plan for a slightly longer than average straightaway leading to the runway to give him a few extra seconds to line up with the runway, to determine the correction for drift, and to adjust to a good rate of descent. Previous experience had taught him that, in the windy conditions, partial, rather than full, landing flaps would give him better control once on the runway and that carrying a little bit of engine power right to the point of touch down would give him better control before touch down and provide insurance in case of a sudden, strong gust of wind. Also, in using the slip method of compensating for the crosswind, he could actually keep the aircraft slightly right-wing down until the right main wheel touched the ground first. Doing that would let him control his drift for that critical second or so in the transition from flying to partial flying at touch down and during the rollout on the runway — the most critical part — and then to just taxiing. In thinking of his plans, once more Pete wished he were flying a heavy jet, which he could be relatively sure would stay on the runway, instead of this light Beechcraft which, by comparison, was like trying to make a leaf stay still in a stiff breeze. But Pete well knew from his own experiences, that big aircraft also have crosswind restrictions as well. No aircraft was immune from restrictions in certain crosswind conditions. Even the heavy returning satellite space ships like Discovery are restricted to landings in which the crosswind does not exceed a relatively light ten knots. And though every student pilot is taught about crosswinds, and many seem to think crosswind landings are a matter of concern only in their small, light airplanes, it continues to remain a factor for all pilots, military and commercial, regardless of flight time and experience because there are extreme conditions that are just to be avoided if possible.

Very carefully, but relaxed now that he was free of indecision, Pete completed his landing check list, doubly checking that the tail wheel was in the "locked" or non-swivel position — that was crucially important! Pete completed his approach turn and hit the straightaway just where he had planned and adjusted the throttle a few times to get the rate of descent he wanted. Now in the straightaway, he lowered his right wing to correct for the wind that, without adjustment, would very quickly push him well off his straightaway and to the left of the runway center line. From this position, it would only take a few seconds for everything to be resolved — a satisfactory landing, an unsatisfactory landing and its consequences, or a safe waveoff. As Pete approached the touch down point, he could sense that his drift correction was good and holding steady even in the rough air conditions, an indication that, at least at this moment, the wind was holding steady. Twenty-seven knots (31 miles per hour) sixty-five degrees to the runway was quite enough to cope with without the additional factor of strong wind gusts to thirty or more knots. But would the wind hold steady and free of gusts and what were the odds that it would hold steady on such a day? "Not good!" thought Pete. "And a sudden gust will negate all the good work so far, creating a messy can of worms in one hell of a hurry."

Approaching the ground, Pete began to ease his nose up to start the flare out in order to make the touch down as smooth as possible. Shortly thereafter, he felt the right main wheel touch the runway with a gentle squeak, followed a second later by the feel of the left main wheel touching the runway. Harry Kinsey yelled, "Great," knowing full well that, as good as the touch down was, it was just the first step and that the next few seconds were more critical than the landing. When the left wheel also touched the runway gently, Pete immediately pushed a little forward on the control yoke, which raised the tail slightly insuring that no sudden gust would lift them back into the air. This move also insured that they would have good directional control right after touchdown while their speed was relatively high. So far, Pete could not have hoped for more: he had greased it onto the runway as well as it could be done; and while they were still rolling at a good rate, they had good directional control. But the far more difficult part remained, one that, poorly handled, could ruin everything and overshadow a good approach and landing.

With the tail a bit high and a little power still on the engines, the slipstream from the rush of air from the propellers provided a good flow of air over the horizontal and vertical stabilizers of the tail section, giving good directional control. The crucial point was coming when Pete would ease off the little remaining engine power that was still on the plane's engines and the decreasing speed would let the tail drop to the runway. As the aircraft's speed slowed, there would be a diminished slipstream to provide rudder force to control the direction; and there would be even a point where the slipstream would be blanked out by part of the fuselage. The weak slipstream would require that Pete get the locked tail wheel on the runway quickly so that the friction on the rubber tail wheel would resist the sidewise force of the crosswind on the tail rudders, or more accurately, the vertical stabilizers. Also, once the tail wheel was firmly on the ground, directional control could be maintained by braking one main wheel or the other — in this case with a right crosswind, it would be the left brake that would be utilized because the tendency is for the aircraft to nose right into the wind. Pete, of course, was well aware of this entire procedure and resolved to make the transition from the tail and tail wheel not yet on the ground to tail wheel on the ground as quickly as possible. He only hoped that the tail wheel was locked as he felt it was.

Having thought about all of the things necessary to land in these challenging conditions, and so far having done everything exactly as he had hoped, Pete eased off the little engine power that remained; and the tail started down. Pete sensed that his heading straight down the runway was good so he pulled back on the control yoke to hasten the tail's move to the ground. At the same time, Pete knew he had to be alert to use either brake should any tendency to swerve be imminent. When the tail wheel touched the ground, Pete waited and was ready for any tendency to swerve. As each second passed and the aircraft slowed further towards the normal taxiing speed, it became obvious that no uncontrollable swerve would develop. Though Pete had done everything right, he felt a bit disappointed that he was not further challenged — particularly when he had thought that he had been so well prepared. He knew that, despite previous and obvious concerns, he had handled the Beech SNB pretty well in one of its worst environments — or was he just lucky? He knew also that, had the strong crosswind been from the left, the challenge would have been significantly greater.

CDR Greville yelled, "Good job!" He was obviously glad to have reached the Martin field, and departed the aircraft quickly with a heartfelt "Thanks."

The ensuing takeoff in these conditions would require some skill also. In effect, Pete would have to do the same things but in reverse order. He would have to keep the locked tail wheel on the ground until the engines had built up a safe takeoff speed to insure he had rudder control. He would also have to compensate for the loss of some lift on the port wing because the fuselage blanked out a part of the wind from the right and hence the lift; to do this, he would ease the control yoke to the right to raise the right aileron a bit and lower the left aileron, thereby reducing the lift on the right wing and somewhat raising the lift on the left wing. Also, there was a strong additional air pressure force against the tail's vertical stabilizers, which was caused by the swirl of air created by the churning propellers, a swirl of air forced down under the fuselage which rolled back up in time to hit left side of the vertical tail surfaces. Further, it was desirable to take off in such a way that, once airborne, he would not drop back and bounce off the runway a bit: there would likely be a little drift immediately after becoming airborne; and a bounce back to the runway, while not particularly dangerous, would put additional stress on the landing gear. Some skill was required, but Pete Larsen was quite familiar with what he had to do, and they were soon airborne.

After becoming airborne and climbing to cruise altitude, Pete started to think about what they had just done. But had he really really done it? Pete questioned whether he had "pulled one off," as they say, by being successful in handling the Beech. Had he really taken an unnecessary risk? He began to think more and more about this on the return flight while LT Kinsey continued his instrument work. "Was this mission really worth the risk?" Surely it was not. One person would have been inconvenienced — that's all. Hadn't he learned a long time ago that, in flying you had to know when to say, 'No!' And experienced pilots knew that there had been serious incidents when someone had tried to make a questionable landing just to accommodate someone. Making that landing in conditions that exceed the maximum allowable crosswind component for the SNB was like playing Russian Roulette. You might get away with it if you were exceptionally skilled — at least for awhile; but sooner or later, you would become a loser. Larsen knew his thinking was logical, and he knew as well as anyone the logic of the Navy's rules, vowing that, for the future, he would definitely lower the odds.

Yet despite his thoughts, Pete wondered if was really as simple as that. "What about those night carrier landings in bad weather and in jet fighters? What about those instrument approaches to airfields when the weather is at the minimum allowable and you have a crew of twelve aboard? Do you avoid them because they can be dangerous if you haven't done your homework properly? And besides, how you disliked that rare Naval Aviator who would 'become ill' aboard the carrier when the going got tough, putting an extra burden on his squadron mates. If not in challenging situations, when do you push yourself? When do you use all of the skills you're supposed to have acquired? Where do you draw the line?"

Right now, this line of thinking led Pete nowhere and created an ambivalence he would try to resolve somewhat later. At the present though, he wanted to rid himself of these thoughts. He was reasonably sure that the line which divides <u>yes</u> from <u>no</u>, <u>black</u> from <u>white</u> was not a thin line; rather it was broad and indistinct much like the daily twilight periods. We are aware when it is light and we are aware when it is dark; but the space in between consists only of gradations. Although learning from all experiences, Pete Larsen tended to look to the future optimistically. At this moment, he was positive about only one thing — a relatively good feeling remained. After tonight's flight, the drink or two he might have would taste a little better, he would enjoy the company of his squadron mates a little more, and he would think the stories coming in the sessions when the hangar doors reopened were more interesting and even a trifle funnier. Today's flight to Martin Airport would no doubt be included as one of the stories that would come up when one pilot relates a somewhat similar incident.

More About Crosswinds

So many things are more complicated than they appear. So it is with pilots' consideration of wind direction and force when landing. Explanations can be presented to make the situations quite simple; but those explanations fall somewhat short of being completely accurate. While it can be shown with vectors what will theoretically happen to an aircraft landing in a crosswind if no compensating action is taken, once the pilot tries to compensate for the crosswind, the forces and the vectors deviate a bit from theory. But for our purposes, let the simple explanations suffice for now, suggesting that the reader search out further information as his interest dictates.

There are two main methods of compensating or correcting for the crosswind. Military pilots, commercial pilots, and other experienced pilots make these compensations almost without thinking about them; but they too must take note of exceptional wind conditions, leaving nothing to chance. One method of compensating is to head slightly into or "crab" into the wind just enough to correct for the drift, thereby keeping the aircraft going down the center line of the runway. The term "crab" is appropriate, because, during this correction, the aircraft is moving sideways like a crab as it proceeds down the runway. Crabbing has its problems, because the aircraft has to be turned into the runway heading just before touchdown, since it obviously cannot land sideways without causing damage to the aircraft. The stronger the crosswind, the more difficult it is to use this method.

A much more satisfactory method is to "slip" slightly into the wind, just enough to keep the aircraft heading down the runway's center line. More precisely, the term "slip" is used to describe a method of losing altitude rapidly and is generally restricted to relatively light aircraft. The "slip" as it refers to crosswind compensation is actually an uncoordinated turn into the wind in which the aircraft's nose is pointed along the runway's center line while a wing is lowered into the wind. The aircraft is actually flying into the wind while its nose is headed along the runway's center line. In this attitude, the main wheel on the right side (in a right crosswind) will touch down first, followed slightly later by the left main wheel. If executed properly, there is no sidewise drift because the nose of the aircraft and the wheels are heading straight down the runway.

While the "slip" method of compensating for crosswinds is preferred — experienced pilots use it without thinking and it is used in most aircraft regardless of size — it too has its limitations. One notable exception to its always being the method of choice is in the Air Force's eight-engined B-52 Stratofortress. Its wings are so wide that, before the "wing-down wheel" would touch the runway, the wing tip could scrape the runway. To compensate for this condition, the aircraft's landing gear can be aligned with the runway while the aircraft itself "crabs" into the wind. On touchdown then, the nose is headed slightly into the wind, while the wheels face the runway heading and continue this way down the runway. Of interest is the B-52 pilot's ability to set the wheels to the runway's direction like Runway 35 (350 degrees), Runway 17 (170 degrees), etcetera if necessary.

BUT! the important thing is to avoid conditions where strong crosswinds prevail. Obviously, anything less than a strong ninety degree crosswind makes landing much easier. In fact, a crosswind of thirty to forty miles per hour at ninety degrees to the runway is quite rare, mainly since there are usually several runways at an airport, providing four or six directions to land in depending on wind direction and force. Even at airports with just one runway, providing just two directions, that single runway is usually aligned with the prevailing wind directions. The aircraft carrier provides the ultimate, with its ability to turn directly into the wind and also with its ability to move into the wind speedily to create favorable wind forces for carrier operations — a decided advantage in wear and tear on the pilot, the aircraft, and the aircraft carrier's equipment when bringing a heavy, high speed aircraft approaching at about 125 knots (143 miles per hour). The carrier's ability to provide strong wind right down the deck makes the touchdown speed considerably less.

The Effect of Winds on Landing Aircraft

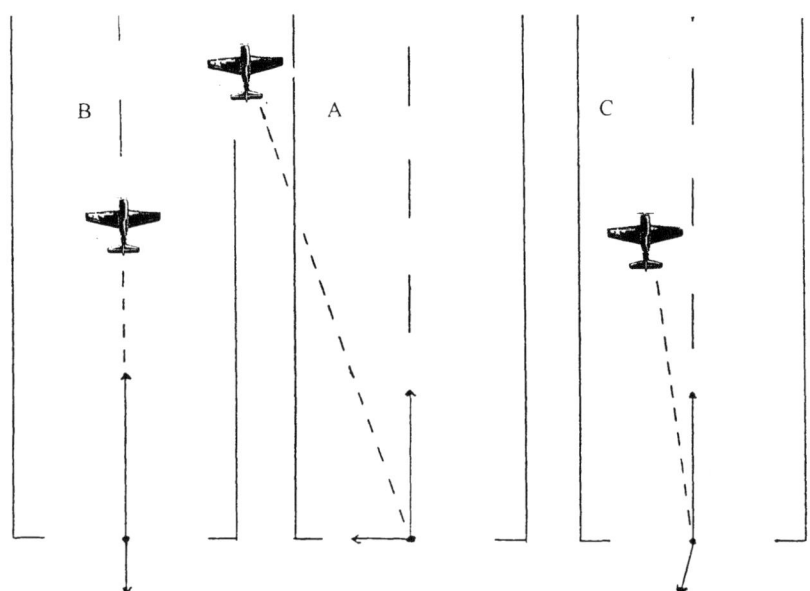

The explanations here are simplified, mainly to stress the principles involved and because lengthy explanations would not add that much. So, assume that the dot on the threshold of Runways A, B, and C is the point at which wind forces, represented by vectors, are applied to the landing aircraft, assuming that the pilot crosses the runway threshold in the middle of the runway width. The vectors represent speeds, with one quarter of an inch equal to 30 miles per hour. In all three instances the aircraft with wingspans of sixty feet are landing on two hundred foot wide runways and are subjected to forces of ninety miles an hour forward, the speed of the aircraft, and winds of thirty miles per hour, which, for illustrative purposes, are coming from different directions. Further, before touchdown, assume that each aircraft is subjected to these forces for a period of three seconds, the time from crossing the threshold until touchdown on the runway (though actual time might be longer). The thin dotted line shows the path of the aircraft, and its touchdown point is where the aircraft figure is at the end of the dotted line — <u>provided the pilot makes no attempt to correct for the wind</u>.

In figure A, the aircraft would touch down off the runway, because the thirty mph wind, exerting a force of 44 feet per second would force the aircraft 132 feet to the left, or 32 feet beyond the left edge of the runway. In figure B, with the wind directly down the runway, the aircraft would touch down right on the center line of the runway. And in C, the aircraft would touch down near the middle of the left half of the runway. In all cases, the touchdown point is an indication of the speed of the aircraft affected by the wind. In A, the aircraft has travelled 396 feet, making its landing speed ninety miles per hour, while in B the aircraft has travelled 264 feet, making its landing speed 60 mph. In C, the landing speed would be a bit higher than in B and the landing point a bit farther down the runway, since the wind is not right down the runway.

What Was "Poor" Chuck's Problem?

What was poor Chuck Raymond's problem? Military pilots face many problems in their careers. Even those who fly for just a few years have their challenges. But LT Charles Raymond's problem was somewhat different, if not necessarily unique; and how he and others faced his problem makes his story worth telling.

It wasn't that LTJG Chuck Raymond wasn't intelligent. He had a bachelor's degree from a state university. He had passed the necessary aptitude tests for flight training, and he handled his assigned collateral duties as the squadron's personnel officer reasonably satisfactorily, if quite a bit short of exemplarily. He did seem, however, to be overly concerned with minutia and what could only be construed as gossip, taking under his wing newly arrived, young officers to give them the lowdown on the squadron and its officers; this was a self-appointed task he seemed to enjoy. In these actions, he seemed singularly different from Naval officers anywhere else. Chuck was nice looking and far from sickly. On the contrary, his six foot four inch height and his two hundred twenty-five pound body radiated good health. He seemed exceptionally confident and reasonably cheerful in most settings, even to the point of affecting superiority and condescension. In fact, how he could be so outwardly superior yet do what he did was a puzzle. So, it was not easy to figure out what his problem was or why he reacted to it as he did. Only one thing was certain: the problem manifested itself in the same way every time the squadron went aboard an aircraft carrier for military flight operations.

About noon on the first day of the *USS Midway's* (CVB-41) departure from the Norfolk Naval Base, en route to the Mediterranean to join the Sixth

Fleet, the flight schedule for the next day's — the first day's — operations was posted on the squadron ready room's blackboard. Only eight planes, consisting of two four-plane divisions, were scheduled for the early morning's (0700) launch for practice dive bombing runs; and there, scheduled as the division leader of the second division of four aircraft was Chuck Raymond's name. Also scheduled were two supernumeraries or alternates who would stand by for emergency pilot replacement. The first supernumerary was LT Bill Klug. Being a supernumerary did not usually result in your having to fill in for someone. Ordinarily, that was very rare. But in this case, although it was not possible to know what Chuck Raymond was thinking, Bill Klug's thoughts were clear in his own mind — and perhaps a bit unkind. Klug was a very conscientious officer, as were practically all the pilots; and he was a very good, very responsible aviator. His immediate thought on seeing the flight schedule was that Chuck Raymond was on it and that, if Chuck's actions held true to form, "Poor" Chuck would become ill and be "grounded" by the medics. In a subsequent thought, Krug figured that Chuck would remain grounded for some time, maybe even the entire time at sea. Klug kept these thoughts to himself, figuring that they were too mean-spirited and beneath him to utter openly and that they might make him seem petty, though he knew that quite a few of the other pilots in the squadron had arrived at the same conclusion, because Chuck's becoming ill was a regular occurrence whenever the squadron was at sea aboard the aircraft carriers.

In keeping with these thoughts, Klug expected to see Chuck's name erased and his own name placed in Chuck's spot before the day was over. The thought was annoying to Klug, because, under the circumstances, Klug would much, much rather be on the schedule than be a back-up for Chuck. But in this case, Klug was wrong. The afternoon passed and then the evening, and Chuck's name was still on the schedule. By 2230, when Klug turned in for the night, there had been no change; and Klug felt that it would be unlikely for there to be a change at this point. Klug started to think that perhaps he was wrong about Chuck and that his thinking such thoughts about him was somehow demeaning and unworthy of himself. After turning in and reading in his bunk for awhile, Klug then fell into a relaxed, sound sleep.

At 0630, there was a loud knock on Klug's compartment door, followed by the message, "LT Klug! LT Klug! Poor LT Raymond became ill during the night and can't fly. You're the first supernumerary!" Awakened from a fairly deep sleep, Klug did not immediately think of Chuck. He thought mainly of getting dressed, getting to the ready room, and of the problems such a late notification would produce. He hoped that, after getting to the ready room, he would be able to yet hear some of the briefing while he donned his flight suit. Although breakfast was out of the question and not really necessary, he knew that was a minor inconvenience compared to what was to follow. In the Ready Room as he took the last item of clothing off to get into his flight suit, Klug realized that he had missed the entire briefing.; but he did learn that a new second division leader had been designated and that he, Klug, would fly on the left wing of the flight leader. And then, as Klug put one leg into his flight suit, over the public address system came the command, "Pilots, man your aircraft!" "What to do?" thought Klug. With no briefing, Klug had no idea where the aircraft were to go and what their bombing target was to be. "Was it wise to fly under such circumstances?" Determined to try to salvage the situation and sensing that going ahead seemed easier somehow than cancelling out at the last second, Klug got his helmet, Mae West (life vest), plotting board, knee pad, etcetera, and joined the others on the flight deck to man their aircraft. Although the events had been very rapid and confusing so far, Klug felt that he could probably piece together much of the briefing as the mission unfolded. He was not prepared for one further surprise that hit him shortly after the group took off from the carrier. No sooner than all of the aircraft had joined up in flight, it became apparent that the flight leader, LT Fred Stamm, was having radio problems when trying to communicate with his flight's pilots on the squadron common frequency. Feeling frustrated and handicapped, Fred Stamm thought it best to turn over the flight leader's responsibilities to someone else. He turned his head to the left to his left wingman, now Klug, forgetting in the confusion of the moment, that Klug had not been in the ready room in time for the briefing. Stamm tapped the top of his helmet twice and extended his open hand in the direction of Klug. This was the standard means of transferring flight leadership by gesture, when the radio was not to be used or, as in this case, there was a radio problem. There was no way that Klug could have assumed the leadership, and so he shook his head indicating "no," all the while thinking how bizarre his assuming leadership would have been. It then dawned on Stamm that he had to make another choice, a choice that then suddenly became unnecessary as his radio problem cleared up sufficiently to lead the flight. As he thought he possibly would, Klug managed satisfactorily, being able to piece together over time the essence of the information he had missed at the briefing and even beginning to relax since that first loud knock on his door at 0630. But, of course, it was not the way Klug would have liked it, preferring, of course, to be on the flight schedule from the beginning and being able to do things in the officially prescribed manner.

The morning's series of events had quite a comical side as told afterwards by the pilots on this mission, some mimicking Klug as he put his first foot into his flight suit. And for several days thereafter as pilots were just starting to put their flight suits on, someone would yell, "Pilots, man your aircraft," evoking a relaxing laughter not usually present when preparing for a mission from the carrier. This humor passed over the head of "Poor Chuck," who, not present when the original incident occurred, had no idea of what was going on. Later, being advised of what prompted the laughter, Poor Chuck did not think it was funny.

But what about Poor Chuck Raymond? After landing back aboard ship from his "fill-in" mission and seeing Chuck, Klug noticed that Chuck did not look ill. In fact, he seemed to be the healthiest looking specimen of a strange or unknown malady that you could possibly find. Certainly, he seemed to have no cold or any congestion as some of the pilots obviously had and were contending with; but his illness, undefined as it was — perhaps a nebulous back problem — deprived him of the exhilaration which the challenge of aircraft carrier aviation provides. He continued to be so deprived for a long time, until the carrier docked at Pier Seven of the Norfolk Navy Base several months later, when his apparent, symptomless, seabourne malady seemed very much improved. Whatever illness or problem Chuck had, he did not evince a lack of self-confidence or evidence a lack of self-esteem as a result of his seemingly feigned "illness;" nor did he seem to have a sense of guilt or shame that his actions might have placed additional burdens on his squadron mates. Had a pilot filling in for him been killed in an accident, it seems unlikely that he would have been affected. No doubt, the same shamelessness that enabled him to feign illness would let him act as if nothing had happened at all. Whatever complaint he described to the flight surgeon must have been convincing — a complaint made all the more believable by Chuck's impressive but false manner of sincerity — because the doctor fell for it.

Although everyone knew what Chuck was doing, his actions were only rarely discussed and then only minimally. Part of this was the result of his not being on the flight schedule during the remainder of the time at sea; so no other supernumeraries had to be used or inconvenienced by sudden wake-up calls, which would further call attention to his problem and be a reminder to all of the other pilots. Also, with twenty-four other pilots in the squadron, having one or possibly two pilots grounded did not present a big problem — especially when all the other pilots wanted to fly at every opportunity. No doubt, "Poor" Chuck was as well aware of this as everybody else.

Any commanding officer anywhere, who saw Chuck's situation, could make an instantaneous and miraculous cure of Chuck's problem merely by raising an eyebrow — and most certainly by saying something to him. Commanding officers just do not ignore such situations; they don't put up with what some would call malingering. To do so would not only be detrimental to mission requirements assigned by higher authority, but it would also reflect badly on the CO's career. One may wonder why someone did not talk to the CO about Poor Chuck — perhaps someone had, but not Klug. There were several reasons that most would not complain: first, in a very strange and unique situation, Chuck's CO would often keep himself off the flight schedule and, in addition, would also be grounded medically himself; next, the other pilots were not severely critical, figuring that, whatever Chuck's situation was, he alone should work it out; further, other pilots always seemed to feel that they had a duty to step in to get the job done — even if leaderless. In fact, the rest of the squadron's pilots did an excellent job, thereby inadvertently seeming to cover up for Chuck and the CO as well. As practically all pilots know, any squadron commander who does not fully accept his responsibilities, set an example by leading and being out in front, and demand the best of his people is a disaster. There is really no other way to put it. He hurts himself, but he also affects many others. Somehow or other, higher authority knows what is going on, perhaps because the problem manifests itself in other endeavors as well, affecting the overall efficiency rating of the squadron and hence its reputation in the air group.

Such actions, noticed by superiors, result usually, and in this situation, in limited careers for both "Poor" Chuck and the commanding officer. The position of squadron commander is much coveted. Not only does it provide an officer with the opportunity to show what he can do leadership-wise, but it is also an extremely important avenue to promotion and positions of increased responsibility. It is unquestionably better for an officer to decline the post of CO than it is to accept it and perform poorly. The question is always there as to how the squadron with such poor leadership can perform in crisis situations. New pilots and other young officers may not always know it, but they are much better off serving under a good, tough, "ass-chewing," no-nonsense, demanding CO who sets an example than under a weak, diffident CO. It is really under a good, tough, demanding CO that young officers grow and mature; their accomplishments approach the true levels of their abilities; and they very often achieve things that they had no idea they could do. The Navy just cannot put up with the likes of "Poor Chuck" and his CO.

In this true episode, the reader may feel that Chuck Raymond was described unjustly or unfairly. And perhaps he was, but the incident raises many questions. All Navy pilots recognize that landing on an aircraft carrier requires the adherence to precise flight procedures and the utilization of some specialized skills. Flying Naval carrier aircraft at extremely low altitudes and at speeds not too much above the stalling speed as the pilot makes his approach turns to the carrier is not everyone's cup of tea. Add the skill required for night takeoffs and landings back aboard ship, sometimes in bad weather, and you find that that too is not everyone's cup of tea. But it is done because it is necessary, and Navy Pilots develop the skills and determination to do it well.

Yet the reactions of Poor Chuck and his CO are extremely rare. Is it an oversimplification to say that they had personality, emotional, or character flaws? And if so, do these flaws manifest themselves in ways other than in flying? Or is it possible that the flaws related only to flying on and off an aircraft carrier? Weren't all pilots as aware of the demanding nature of landing on an aircraft carrier as Poor Chuck Raymond and the CO? Were the actions of Chuck and the CO calculating and scheming or could they not help themselves? Did they lack courage or self-respect? Did they have strong phobias with which they could deal in no other way, however agonizing it was to them? How could they lack confidence in their carrier flying skills, yet impose on others, portraying a feeling of confidence as if nothing were amiss? What about the inherent obligation of each to carry his load or responsibility and not burden others? Were their actions their only means of surviving, thereby perhaps being more psychological than anything else? Could they be trusted in an emergency or to carry out a difficult military mission? And what of LT Bill Klug's response, stepping in while knowing fully "Chuck's Problem," and making the most out of a botched situation — with no preparation time? Was Klug wise to go ahead with the flight without having sat in on the briefing? Did he pose a danger to anyone or object? With most pilots, the desire to try to make things work is much stronger than the thought of cancelling out.

In retrospect, it would be interesting to hear the diagnosis of a psychiatrist or clinical psychologist based on the brief descriptions of the events described here. No doubt, some broad generalizations could be made. But most likely, they would have to delve much more deeply into the specifics of human nature in these events. One thing is sure: these pilots can be of no use to Naval Aviation. To tolerate such behavior makes no sense, because whatever caused their behavior is likely to prevent recovery. Such an attitude is not to be unkind or unsympathetic. There may be a place somewhere for people like these, but it is not in Naval Aviation. Stronger, more stable people are needed for the requirements of Naval Aviation.

The Judge's Embarrassing Mistake!

CDR Henry "Hank" Morgan was one of those guys who would stand out in a crowd, no matter how bright, accomplished, or distinguished the crowd. And there were quite a few pilots in his squadron and the other squadrons at the Willow Grove Naval Air Station who were working in successful, worthwhile careers despite their having given up four or more years of their lives while flying for the Navy and now being in their early to mid thirties: partners in accounting or law firms; associate professors; high school administrators and teachers; company officers; even a couple of pilots who, after undergraduate education and a period of years in the Navy, were now in their last years of medical school at the country's leading universities. But CDR Morgan, very intelligent, courteous, and sincere, first in his law school class, a partner in a prestigious law firm, and later to become a distinguished judge, was endowed with handsome good looks, a taller than average frame, and a confident, self-assured manner which enhanced his effectiveness. Now at about forty and after two tours of active duty, one having been in a squadron recalled to active duty during the Korean War, he had been flying for nearly twenty years; and he was presently serving as the commanding officer of a Naval Reserve four-engine SP-2E Neptune patrol and anti-submarine squadron. Along with his other qualities, his presence and manner were such that others quite naturally deferred to him. If occasionally he seemed imperious or impatient with others, it was more a matter of "not suffering fools gladly" than a matter of his being intentionally rude. Most people who knew him sensed this, realizing that he approached everything seriously and professionally, paying attention to every important factor to ensure that all flights and other squadron business were responsibly performed. He seemed to inspire in others the same desire to do things right: and they wanted to be a part of his undertakings, knowing full-well, that they would have the satisfaction of a job well-done. Not devoid of humor, and self-deprecation, CDR Morgan unquestionably was well-liked; but more, he was well-respected. So, even if he was not yet a judge, he was often referred to as "the judge." While none could predict his ascension to the bench, likewise no one was surprised when he became, as he was called earlier, "the judge."

Hank Morgan's squadron had just completed twelve days of its annual fourteen day period of active duty at the Naval Air Station, Brunswick, Maine, and was making preparations to return to the Willow Grove Naval Air Station. The active duty period had been excellent. Having achieved most, if not all, of the training objectives he had hoped his squadron would accomplish, including an average of fifty-five hours of flight time per pilot and having demonstrated to his corresponding active duty squadron people at Brunswick that his was a competent outfit, he and his squadron mates, possessing a good feeling of accomplishment and time well-spent, looked forward with anticipation to getting home and getting things wrapped up at Willow Grove and then getting back into the flow of their civilian careers, family life, and other responsibilities. But first, as was nearly always the case, their two week period of active duty would be extended by two days at Willow Grove, the two days which would ordinarily constitute their monthly weekend drill days. During these days there would be additional flights and training; but there would be time also to wrap things up, to file reports, complete paperwork, and have a final meeting for all hands late on Sunday afternoon. The meeting would be a good time to review progress and acknowledge the contributions of all hands: it was a time that just naturally seemed to bring everyone together and to instill squadron spirit and pride — and, indeed, they had much to feel good about themselves; their active duty endeavors would become frequent topics of conversations for some days to come, not only at home but also at their places of employment. They had had some interesting, satisfying experiences, involving long days; and their civilian co-workers would naturally be interested in hearing about them because the work was quite different from work "at the office."

But first, they would have to return to Willow Grove. The five SB2E Neptunes in the squadron, each with a crew of twelve, would proceed independently to Willow Grove, while the rest of the squadron, another seventy men, would be airlifted back in Navy transport aircraft. The pilots in the five Neptunes would file instrument flight plans, even though the weather both at Brunswick and Willow Grove was good. The filing of instrument flight plans would insure two things: that even if there was some bad weather en route or if the weather should deteriorate, they would be under the positive control of the Federal Aviation Administration's (FAA) controllers while they were flying in their own "spot" or slot in the sky; also, they would be flying in simulated instrument flight, the same kind of flight they would make in bad weather conditions when they could see nothing but the inside of clouds. As noted, this kind of flight would require the filing of "instrument flight plans" with the FAA, enabling the pilots to practice and to gain additional instrument flying experience, utilizing the last two hours of flight time as they returned to home base. After filing his instrument flight plan, Hank joined his crew at their assigned aircraft, where each member of the crew looked over and inspected his own areas of responsibility. The preflight inspection completed, everyone climbed aboard; and Hank Morgan taxied the plane along the taxi strip to the warm-up area, where he would go over the detailed check-off list required of the SP2E and run up his two big reciprocating engines, leaving the jet engine start-ups until last when they might reasonably expect to receive their FAA-approved flight plan. The instrument flight plan from FAA would assign the specific airways (the routing) to be flown, the altitude to be maintained, instructions following takeoff to reach the airway and altitude, and any special communications instructions right after takeoff. On days when air traffic is heavy, the wait for an instrument clearance from FAA can be long; but this was Saturday afternoon, the fifteenth day of their sixteen day training period, and the weather was good. In these circumstances, Hank knew that the wait for a clearance should not be long. Barely had they completed the check-off list and started the jet engines when the clearance came through. The procedure required that the pilot or copilot first copy and then repeat the clearance back to the control tower exactly as it had been given to avoid any misunderstanding. With everything in order and given clearance for takeoff, Hank taxied onto the duty runway and lined up; with brakes holding, he advanced the power on the reciprocating engines to forty-eight inches of manifold pressure. Then with copilot Bill Olsen placing his left hand behind Morgan's own hand to insure that the throttle handles did not slip back, Hank advanced the jet engines to one hundred percent of RPM and released the brakes, at which time, he advanced the reciprocating engines to the full power of fifty-four inches of manifold pressure.

Advancing down the runway, the power of the two Wright Radial R3350 engines and the two Westinghouse Jet J-34 engines was quite impressive, their roughly twelve thousand horsepower providing increasingly rapid acceleration. And just after becoming airborne when Hank raised his landing gear and eased his flaps up, the aircraft's acceleration was even more rapid, seeming almost the result of a rocketlike boost. Taking off in a southerly direction and climbing toward his assigned altitude, preparing to join the airway east of Portland, his flight path naturally took him over the ocean, Brunswick being so close to the south shore of Maine. This part of any instrument flight is a busy time for the pilots as they try conscientiously to adhere very strictly to FAA's instructions. It is a time of concentration, particularly until the assigned altitude and airway have been reached and the aircraft has been established in its own special piece of the sky and the power settings and cruising speed have been arrived at and the jet engines have been closed down till next required for landing at Willow Grove or perhaps in an emergency. But, as in any complicated or sophisticated aircraft, the pilot must be alert to a number of other things: power settings, engine performance, radio communications, other traffic, emergencies, etcetera. Once established at altitude and on course and speed, the pilots have a little more time to relax a bit, sitting back and monitoring dials, maintaining their navigational position and assigned altitude, and observing the aircraft's performance.

But, as you might expect, it was just in the middle of this very busy, demanding time requiring maximum concentration that Hank Morgan and his copilot sensed that something was amiss. "What was that something, an unfamiliar sound muffled in the ever-present engine noise?" Did they sense peripherally some minor commotion and movement in the aft section of this ninety foot long airplane? Just moments after their becoming aware that there was something unusual happening, Ordnanceman Weathers called the pilot on the intercom: "Captain, our bomb bay doors have crept open a bit — about four inches." Damn it," thought Morgan, while Olsen thought the same, both at first having only a vague idea of the situation and its possible cause. But it really took only a few seconds for both of them to focus on the problem, what the likely cause was, and what it could and would likely lead to, though neither of them had ever experienced the problem before. However, to Weathers, Morgan replied calmly, "Roger," at the same time praying that the doors would not open wider.

Morgan and Olsen quickly determined that they did not have a system-wide hydraulic problem, deducing then that, if the cause of the bomb bay doors' opening a bit was what they felt sure it was, there were four things they could be sure about their situation: first, there was very little, if any, concern about the safety of their flight; second, their Bomb Bay Door Control Handle in the cockpit would be useless in closing the doors; third, that they knew only too well, that Morgan and perhaps his copilot, but no one else, had caused the problem; and fourth and most importantly, that the bomb bay doors were likely to creep open somewhat farther. The pilots knew something else, which most, if not all, of the crew did not know; it was something that would make their mistake all the more embarrassing — even painfully embarrassing in a way. Besides being ninety feet long, the Neptune's fuselage was not very wide, despite its one hundred five foot wingspan; and except for the ordnance station where an average-sized man could stand, it was impossible for anyone to stand erect. In addition, the aircraft's wing extended right through the fuselage, bisecting the interior and making it difficult to get from one half of the aircraft to the other and requiring that anyone changing locations crawl over the wing on his stomach. Furthermore, the aircraft was jammed full of electronic and ordnance equipment. All of the features, while not disadvantages in pursuing military missions, made it a very poor transport aircraft, as it was now being used, to carry men and their luggage back to home base. Well-designed for its military mission, the aircraft's chief disadvantage when used as a transport was that there was really no space to place or store personal luggage; and each of the crewmen had at least two pieces of luggage.

With so little room to store luggage, it was common and accepted practice to place luggage right on top of the closed bomb bay doors before takeoff. So now, the bomb bay doors in their slightly opened condition became a concern to everyone. Would the doors creep open farther? Could and/or would any of the luggage drop out? If so, how much of the luggage would drop out? In the uncertain circumstances at this time, an attempt by a crewman to rearrange the luggage was out of the question — it was just too dangerous!. Not only was such an attempt impossible; but also, with the exception of a few pieces, any luggage so rescued would have no place else to be stored. It would be better to leave things alone and hope that the doors would not slip open wide enough for any piece of luggage to drop out. Would a quick return and landing at Brunswick salvage things? But events quickly answered all of their questions and dictated their course of action, as the doors crept open a bit farther. It was obvious that they could not proceed with their instrument flight plan because, if the doors opened farther, they could possibly drop twenty-four or more pieces of luggage en route over populated areas. Morgan had no choice but to call the FAA to cancel his instrument flight plan and remain over the ocean until he could determine how far the bomb bay doors were likely to open. Shortly after canceling his flight plan, the full extent of their problem became obvious as the first small piece of luggage slipped out. It was followed shortly thereafter by another piece as the ever-widening doors opened farther. Another slipped through. Then another. Then another and another until, with the doors now opened a full two feet or more, every blessed piece of luggage had fallen through the slot and into the Atlantic Ocean, never to be recovered. One can only imagine what a curious sight the fishermen below saw. It must have seemed that some unseen target was being bombed by suitcases. One may likewise imagine both the dismay and wonder of the crewmen as they saw their luggage, with who knows what clothing, valuables, and presents, slide through the open bomb bay doors.

That this incident should have occurred to Morgan was ironic. Here was a good, competent man who, months earlier, while flying the SP2E Neptune miles at sea, experienced an engine fire and calmly shut that engine down, using the recommended procedures and keeping his twelve man crew confident of his abilities. So, why did this strange event with the bomb bay doors take place now? How did it happen that on thousands of similar flights during which luggage had been placed on the bomb bay doors there seemed to have been no similar or reported incidents? No one knew of any such happening, even though all SP2E pilots knew it could happen and must have happened at some time. Any Neptune Patrol Plane Commander knew and CDR Hank Morgan was no exception. But Morgan's knowing why the problem had occurred was no consolation to him. In fact, it made him more upset. It was not that he wished the problem had been caused by some other, perhaps more serious, hydraulic problem. It was just that he felt so embarrassed because he knew that, with all of the details of the period of active duty for training in his mind, he had overlooked a very small but very important detail of his preflight aircraft inspection. As part of that inspection, the pilot is required to check the position of a small two-inch lever in the bomb bay called a "Manual Check Valve." Designed as a safety measure while the aircraft is on the ground, it is a safety measure for someone who is standing on the ground working between the open bomb bay doors. In effect, it disables the cockpit bomb bay door lever, precluding the inadvertent closing of the doors. When the worker is finished, he repositions the lever. During his preflight inspection, the pilot is required to check the position of this lever since the bomb bay doors are normally in the "open" position when the aircraft is on the ground. Quoting from the pilot's handbook, "The valve handle can be placed in the vertical (closed) position to prevent inadvertent closing of the bomb bay doors when ground operations are being performed in the bomb bay. Otherwise, the valve handle should be left in the horizontal (open) position for normal door operation." In the horizontal position the lever lies along the direction of the hydraulic line, much as most domestic water or gas line shut-off valves are positioned to indicate that the line is open, so the pilot can tell at a glance in which position the lever is. And now what is obvious is that if this lever is not in the right position, the bomb bay doors can creep open in flight, and the pilot has no way to close them. Perhaps, if a crew member were familiar with the system and could have acted quickly enough, he could have done something. But in the confusion of the brief time it took for the doors to start creeping open and a potentially dangerous situation to develop, it would have been foolhardy for anyone to approach the doors without a seat pack attached to the parachute harness he was wearing; so it is doubtful that the problem could have been corrected in time.

Later, back at Willow Grove, Hank's feeling of letting his crew down took hold. He was plainly embarrassed and felt his reputation as a good pilot, his good record as a veteran of the Jeep or Escort aircraft carriers in the Korean War, and his professionalism had been tarnished. But if Hank felt that way, his copilot and other crew members did not. They knew him as an exceptionally good pilot and a thoroughly good man. And Hank too was a very logical and thoughtful person; his rational mind convinced him of several things, helping to overcome any real discomfort he might have felt. After all, no one was hurt or even put in danger. The pieces of luggage and their contents, whatever they were, could be replaced monetarily at least by having each man submit a claim for things lost in the line of duty.

Privately, Morgan confided that the idea briefly crossed his mind that, to avoid calling attention to this incident, he would have preferred that no one submit a claim; for each claim with its explanation of loss would only reinforce and prolong attention to his embarrassing mistake. But, however much he may have wished to submerge the incident, he was too big a person and his honest nature and interest in his men precluded any such thought and made him insist that they all submit claims. The crewmen were not responsible, and they were entitled to reimbursement. If CDR Hank Morgan was worried what his superiors would have to say, he needn't have given it a thought. Nary a word of criticism came from above, perhaps because those who knew the Neptune also knew that that very incident could have happened to them. In another sense, one that he did not think about at the time, Hank Morgan gave each of his crewmen a bonus: An experience that each man could relate for years to come, no doubt, each telling being an embellishment of the previous and each being more humorous than the last.

If the mistake of CDR Morgan, competent and experienced that he was, was embarrassing, imagine what a young rather less-experienced pilot felt after his own lapse. At the Naval Air Station Quonset Point, Rhode Island, a young Navy Pilot taxied his AD Skyraider out to the warm-up area near the takeoff end of the runway to go over his takeoff check-off list and check his engine. He revved up his Wright R-3350 engine to check his magnetos, the electrical generators which produce high voltage for the sparkplugs. Reaching the desired power setting of thirty inches of manifold pressure and twenty-two hundred RPM, he checked both banks of the magnetos for the telltale drop in RPM's, which, if excessive, would require him to take the aircraft back for work. On this flight, however, the drop in RPM's was insignificant, indicating he had excellent engine power for takeoff. After the magneto check, he carefully went over the rest of the takeoff checklist to insure that everything was set as it should be. Satisfied that everything was in order, he radioed the control tower and asked for and was granted permission to take off. Granted permission to take off, he taxied onto the runway, revved up his engine, started his takeoff run as he pushed his throttle forward to full takeoff power. The engine was perfect, pulling his Skyraider down the runway with rapidly accelerating speed.

Perfect so far? Not quite! The pilot still had the wings of this aircraft carrier type airplane in the FOLDED POSITION. At what precise point he became aware that his wings were still folded is conjecture, but his astonishment at finding himself airborne and discovering his condition is not. What to do? But the Skyraider, not encumbered with heavy bombs and rockets, was not particularly heavy and climbed to several hundred feet — this with the wings folded. In the wings-folded condition, the aileron controls were ineffective, leaving only engine power and tail surface controls — at the very best, under the circumstances, an imminent crash landing, something that occurred moments later. Remarkable as this takeoff was, even more remarkable is the fact that the pilot survived the crash and was able to walk away from the aircraft.

Aside from the reasons for this incident, that being left for the accident investigators, the interesting part is how the aircraft became airborne at all. The power of the three thousand horsepower engine, described more fully in other incidents, was a decided factor; but the very important additional factor was the amount of wing area on each side of the aircraft which extended out to the point where the wings fold. In other words, the number of square feet of the remaining horizontal portion of the wings had to be sufficient to carry the weight of the aircraft. For example, a wing designed to support a wing load of fifty pounds per square feet would have to have 200 square feet to support 10,000 pounds. In the case of the Skyraider, a plane designed to carry its own weight in ordnance, the level area between the fuselage and the wing-folds, was considerable; and while hardly recommended, the adequacy of that wing could probably be verified in another flight to replicate the first.

P-2 NEPTUNE (Courtesy of Lockheed Martin Corp.)

Open the Hangar Doors

F9F-6 COUGAR

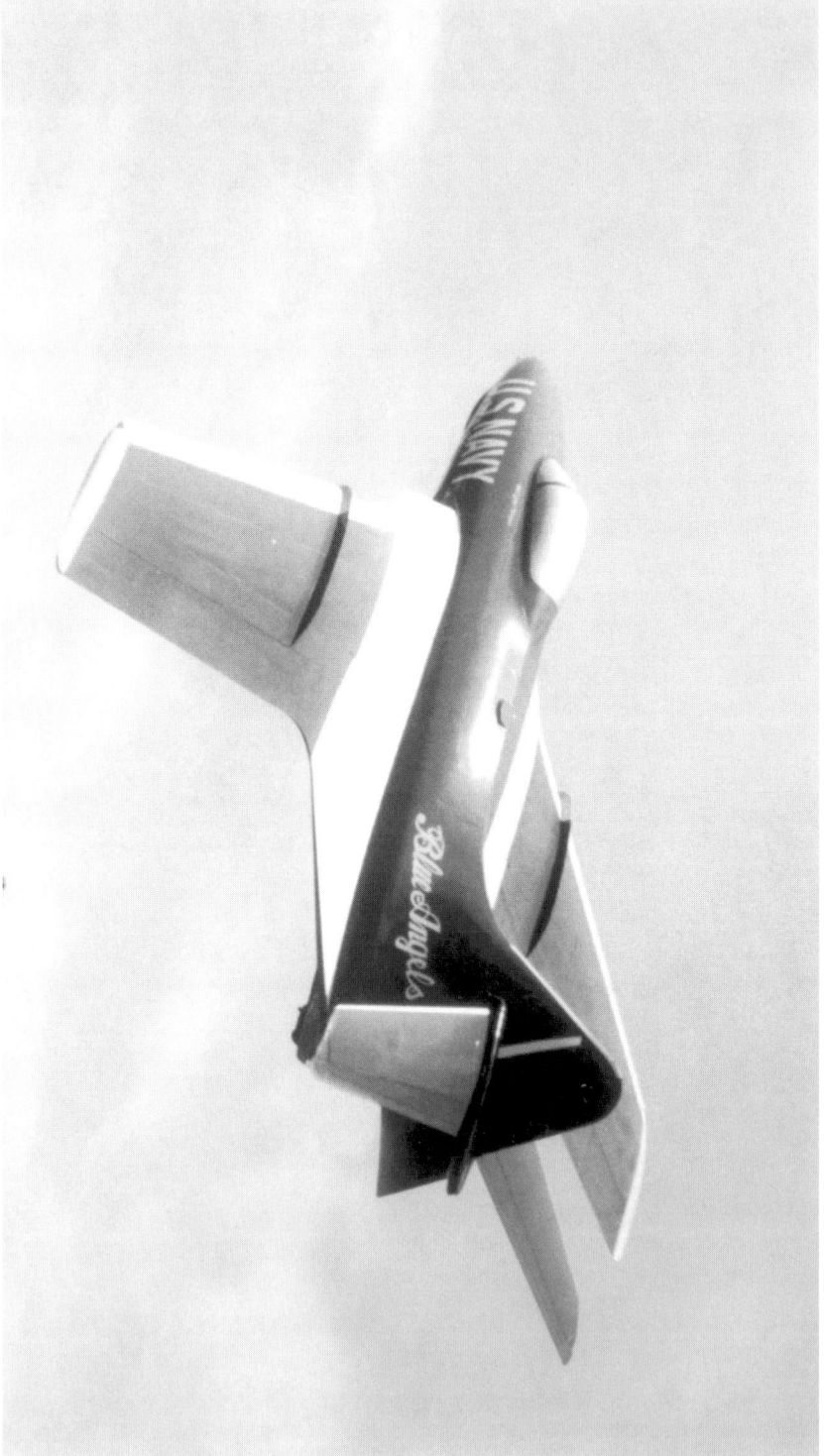

(Leute Photo)

Open the Hangar Doors

(Navy Photo)

AD SKYRAIDER

Open the Hangar Doors

SNB BEECHCRAFT

(Leute Photo)

Open the Hangar Doors

(Navy Photo)

A-6 INTRUDER

Open the Hangar Doors

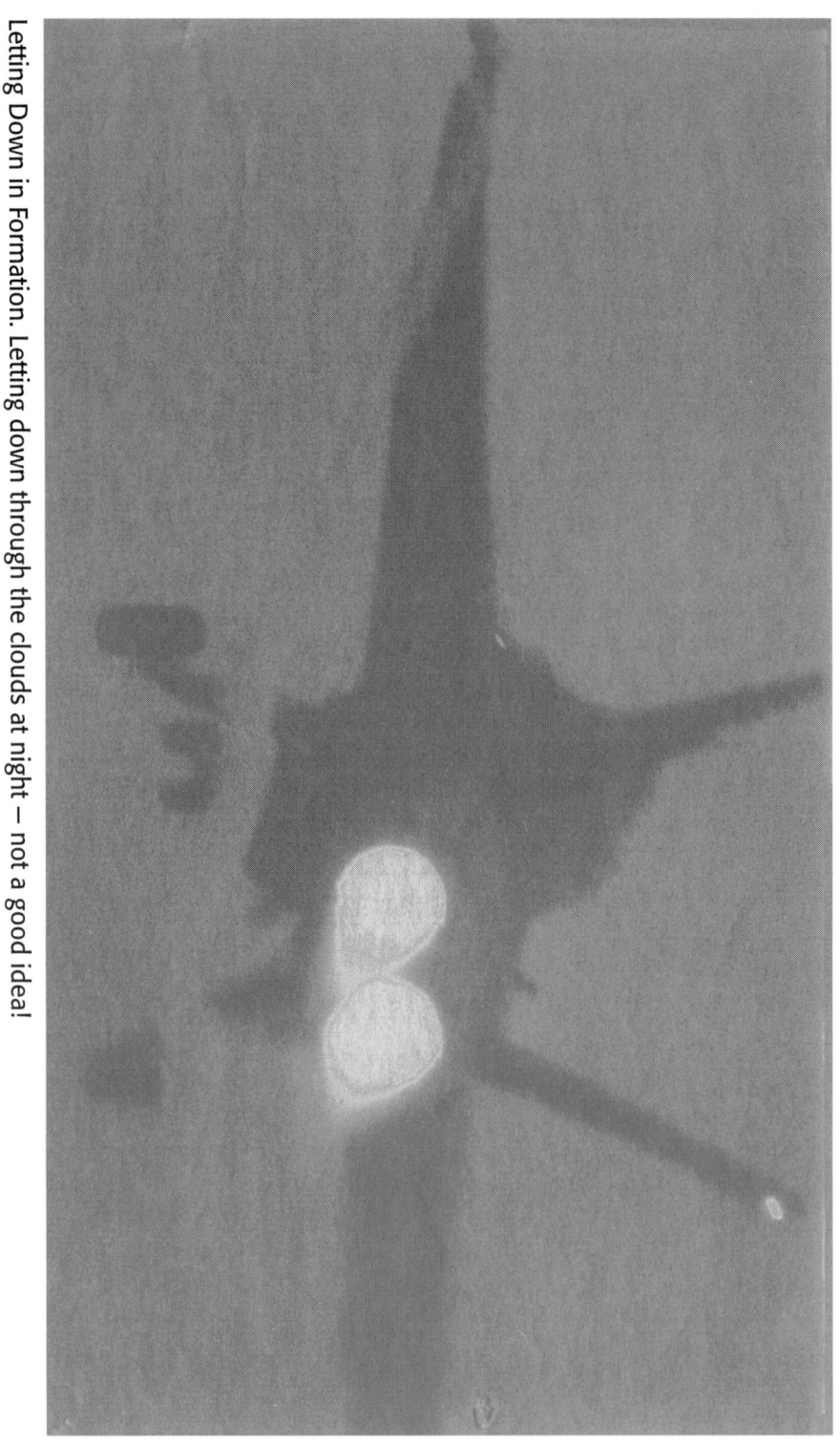

Letting Down in Formation. Letting down through the clouds at night – not a good idea!

Open the Hangar Doors

(Navy Photo)

F-8 BEARCAT

Open the Hangar Doors

A-7E CORSAIR II

(Courtesy of Vought Aircraft Industries)

Open the Hangar Doors

Part Four: Emergency Bailouts

Emergency Bailouts — High, Low, and Other

That parachuting, or bailing out, of a military aircraft is, or should be, rather simple and easy might be a natural assumption. Such thinking would not seem unreasonable. After all, many people participate in the sport of "skydiving." And the pictures of large numbers of paratroopers bailing out in wartime or in military exercises make the case for something simple and relatively safe, even if exhilarating for the participant. And while no professional military man would say that parachuting contains no risk at all, he would know that, using all the correct procedures, the risk becomes minimal — so low in fact that the paratrooper is willing to jump and jump and jump. And the military services have special skydiving teams with excellent safety records like the Army's Golden Knights and the Navy Seals' Leap Frogs, who put on very impressive exhibitions at the Army-Navy Game and on other occasions.

So, considering the foregoing, the uninitiated would probably be surprised to discover that bailing out of an operational or combat type of aircraft not designed or configured for paratroops is neither simple nor quite safe. It is true that the equipment and procedures developed over the years have made egress from high-speed, high-altitude jet aircraft not only more and more safe — but also more and more sophisticated. Even in the days of slower aircraft, jumping out was no sure thing. So, no matter what, bailing out of all military aircraft that are not designed for parachuting has to be considered an emergency, an emergency often exacerbated by rapidly-deteriorating conditions. For single engine, single piloted aircraft traveling at high speeds, there is really no such thing as just jumping over the side. There would be no way to avoid being struck by the tail surfaces in a collision which would make the leading edges of those surfaces knife-like and almost certainly kill or seriously injure the pilot. The solution, of course, is that the pilot must be ejected from the aircraft at a very high ejection speed and with a force so powerful that he be hurled high and safely away from the tail surfaces. Such an ejection requires sophisticated equipment to do the job, and also requires that the pilot adhere to certain pre-ejection procedures, which he must have the patience and time to do!

The means of ejecting the pilot is the appropriately-named "ejection seat system," a system which has been refined over the years, but which now depends on a two-phase use of explosive cartridges and rockets, with the cartridge firing first and the rockets firing an instant later to keep the instantaneous "G Forces" on the pilot's body from becoming excessive. But even if everything works as it should, the pilot must go through procedures to avoid being injured as he is ejected: his arms and legs must be held close to his body and seat and he must utilize a face curtain to protect his face and torso. Otherwise, his arms and legs could be very seriously injured as they come in contact with the overhanging edges of the cockpit. Fortunately, except in extreme emergency, the system design requires that, after the pilot's legs and arms are in position for safe ejection, the pilot must reach up over his head, grab the handles on the face curtain, and pull the curtain all the way down to his waist to engage the firing mechanism.

Under favorable conditions, if they ever are favorable, the pilot has time to do everything as designed. But it is the nature of military aviation that combat situations often result in damage to the aircraft, in which case the ejection system might malfunction or the aircraft might be tumbling uncontrollably. There could also be fire; and depending on the amount and intensity of the fire, the pilot might have no choice but to try to eject immediately. In such a situation, as well as in many other known situations, the pilot may not have had time to do the preliminary steps and may have to be ejected right through the Plexiglas cockpit canopy. Perhaps a less serious situation requiring egress and using the equipment and procedures as specified would occur when the fuel has been exhausted or the engine has flamed out — provided either occurs at a reasonably high altitude, an altitude allowing sufficient time to use the equipment to best advantage. In either case, there is no engine power and the aircraft is descending, although an engine that flames out can sometimes be started; no midair refueling can be accomplished in those circumstances. Lacking the ability to restart an engine or obtain more fuel in these situations, and knowing that he should not ride a jet aircraft below 2,000 feet because the likelihood of being able to land satisfactorily is extremely limited, the pilot is likely to have time to steer his aircraft to a lightly populated area and also to adhere to the procedures which promise the best chance of making a successful bailout. With a solid and thick cloud cover below him, the pilot's ability to do both well is much reduced.

It is not too difficult to dream about all sorts of parachute equipment and devices, in this day of scientific progress in so many areas. In earlier times of slower moving aircraft, when the engine failed on takeoff, the pilot, following recommended advice, would land straight ahead, hoping he could set his aircraft down easily into the wind so that his speed at touch down would be as low as possible. With the advent of high-speed jet aircraft, such a maneuver might still be possible; but touch down speed would most likely be too fast to avoid serious injury or death. So naturally, some parachute device which would enable the pilot to eject himself while his aircraft was still rolling down the runway at high speed was considered. There followed an ejection seat with a special parachute that would allow the pilot to eject in such conditions. The first such chute was called a "zero lanyard" parachute, and was later called a "zero, zero" parachute, the first zero referring to altitude, and the second to airspeed. This kind of parachute required not only an ejection seat to get the pilot well clear of the aircraft but also a means of blossoming or blowing open the parachute to allow a reasonably safe descent. Nowhere is such a parachute more needed than in aircraft carrier operations. The approach patterns to aircraft carriers require low altitudes and airspeeds not too far above stalling speeds. Sometimes, a pilot making a landing approach to the carrier may experience engine failure or stalling, resulting in the aircraft's starting to fall or roll over — or both. While the degree to which the roll has progressed is a very decisive factor, ejection systems are being considered, which, when the aircraft is upside down in flight, will effect a turnaround of the ejection system from going down to going up in the space of a hundred feet. But no matter what, there is no system which can protect everyone all of the time because split-second decisions followed by immediate action may be required; and not everyone is capable of realizing the need quickly enough and then deciding and acting — all in a second or two!

High-Altitude Bailout: Not a Piece of Cake

It was just about midnight, nearly forty minutes since the Pennsylvania Railroad train pulled out of Philadelphia's Thirtieth Street Station en route to Pittsburgh, Pennsylvania. Lt Greg Burns, finally seated in a Pullman car after insuring that train arrangements were in order and that the funeral casket was properly stored in the baggage car, was now reasonably able to begin to think about recent events, events which had him, at present, escorting the body of his good friend LT Tom Trundel to an undertaker, who would meet the train in Pittsburgh and handle the arrangements for the viewing and burial agreed to by Tom's wife and family. Greg reviewed in his mind the events leading up to the problem which led to Tom's bailing out: the extremely urgent bailout from an F9F-6 swept wing Cougar jet at an altitude of thirty-five thousand feet or more; his descent through thick clouds; and his landing in the Atlantic Ocean some twenty-five miles off shore where his body was recovered about an hour or so later.

Tom had been a good pilot, a bright careful pilot with about thirty-three hundred hours of flight time, a higher than average amount of flight time for a thirty-three year old Navy pilot of his rank. He had never had an accident. Why then had this good pilot with lots of flight time, who knew the risks of bailing out at high altitudes and who seemed to do everything right in this instance, lost his life? Greg Burns thought back to the morning of May fourth, just two days ago, when, during the regular morning safety meeting, he and the other squadron pilots could see Tom taking off at about 0810 on a weather observation flight, something routine when the weather could affect the day's flight schedule and operations. The weather flights were scheduled for each pilot on a routine, rotating basis, the assigned pilot then flying one of the experimental squadron's various types of jet fighter aircraft to which he was regularly assigned. After taking off, the weather pilot would head to a nearby radio beacon a few miles from the coast from which he would climb to altitude in a restricted climb space. He would climb in a pattern which would have the pilot climbing in an area off the Atlantic Coast on three different headings. The first heading would be 060 degrees, the second 180 degrees, and the third so that the heading would lead back to the radio beacon from which the climb began. On each of these headings, the pilot would maintain his heading until he had climbed 10,000 feet, with the result being that, by the time he had returned to the beacon, he would be at 30,000 feet or more. Usually, upon returning to the beacon on the third heading, the pilot would be "on top," that is above the clouds. If not, he would take up a heading of 060 degrees again until he was "on top." Although the weather, no matter how bad, would rarely result in cancelled flights, Tom's report on weather conditions would help indicate whether changes in the types of flights already scheduled should be modified. It almost seemed that poor weather was preferred because it would enable the squadron to practice its all-weather, twenty-four hour operational capability.

Climbing at three hundred seventy-five knots (432 mph), the prescribed climb speed at ninety-eight percent of engine RPM, Tom was on top of the clouds at thirty-seven thousand feet in just a few minutes. What he observed during his climb and while he was "on top" of the clouds no one ever heard because he was to report that information on landing some time later. Tom, having made his observations, reported to base that he was over the navigational beacon at high altitude and just about to begin his letdown when he was heard to say frantically over the radio, "I'm getting out of here!" That's all; there was nothing more! The reason for his sudden bailout, like the anticipated weather report, was never heard, leaving only estimates of situations which would necessitate his bailing out. Most of the squadron's pilots, like pilots everywhere, placed two conditions foremost as the likely cause of any bailout: fire or structural failure — after all, to bring an aircraft down in war, bullets and shells are employed to bring about structural damage, damage which may bring about fire as well. Other possibilities included disorientation like vertigo or stalling and spinning down into the clouds where the lack of visual references would immensely complicate both. Still another possibility could have been the loss or lessening of flight control effectiveness, though that seemed unlikely. Although fire in the F9F-6 Cougar was unknown to squadron pilots and structural failure, without a cause, in a rugged carrier aircraft would have been extremely unlikely, and while Tom's flight experience would seem to rule out stalling, spinning, and disorientation, there was something there that made Tom decide to bail out suddenly. But what?

As to the problems which Tom confronted in ejecting from and after he ejected from his F9F-6 Cougar, all were generally known, though none was specifically known in his situation. But it had been well-established that he was a savvy guy and did the required things very well. At what altitude he opened his chute was not known; but since the clouds were solid from seven hundred feet up to thirty-five thousand feet, it seemed likely that he broke free of his ejection seat and opened his chute soon after ejecting. As prescribed, he had disconnected his oxygen mask from the main system and connected it to his emergency bail out bottle attached to his parachute harness — and, based on things he was able to do later, the emergency system apparently worked as designed. Since the temperatures at 35,000 feet could have been as low as fifty degrees below zero, the extent to which these frigid temperatures had sapped Tom's strength during the early phase of the bailout could not be known. If Tom pulled his ripcord shortly after bailing out, he was exposed to the severe cold much longer. Specifically known, however, was the fact that, as he was getting close to the ocean beneath him, he did have the strength to unfasten his shoulder/chest harness and his seat harness as prescribed, so that at the proper time he could slip out of the harness and into the water. Such a procedure would lessen his chances of becoming entangled in the parachute while in the water. Additionally, Tom had inflated his Mae West life preserver. When found in the ocean, Tom was still somewhat entangled in his parachute; and the Navy doctor listed his cause of death as "exposure and/or drowning," a reasonable finding under the circumstances since he seemed to have no injuries caused by the firing of his ejection seat. Although no one will know for sure, it is of more than passing interest to wonder what his fate may have been had he landed in a field where the additional ocean hazards of drowning, parachute entanglement, and heightened exposure would not have been factors. And what if he had been wearing his "poopy suit," the required exposure suit when flying over water whose temperatures were below sixty degrees Fahrenheit.

As Greg mulled over these things, the Pullman porter made up his sleeping berth, and Greg changed his thoughts from Tom's incident to the present; and Greg could not help but think how the mundane can intrude on very sad or very serious thoughts. At this point, he thought it might be wise to get into his lower berth, take his uniform off, and try to get some sleep. Though young, agile, and fit, Greg found that undressing in one of these berths was a frustrating challenge; and he wondered how older, less agile people coped — or did they just sleep with their clothes on, an option Greg did not have since a neat, unmussed uniform would be required for the days ahead. When he finally lay flat and pulled his covers up, he thought that he would be relaxed and drift off to sleep. But no! Though somewhat relaxed, he remained wide awake.

Again, Greg thought back to squadron events the day of the accident and to the two days after the accident. There had been no special safety meeting convened after Tom's bail out to review recommended procedures for high-altitude bailouts. Those procedures had been reviewed many times before, so it was generally known that certain actions would help lessen the risks. For example, the considered judgment was that, after bailing out at high altitudes, free-falling or not opening the parachute until reaching an altitude of about twenty-two thousand feet would provide several benefits: the immediate high opening shock of the parachute at such an altitude and such a high speed would be lessened in a few seconds as the pilot's free-fall gradually reduced his horizontal speed at bailout; the risk of oxygen starvation or anoxia would be reduced, particularly if there were some malfunction of the bailout oxygen bottle attached to the parachute harness; and the risk of exposure to the extreme cold of high altitude would be quickly reduced by the speed of the descent through the cold air prior to the pilot's opening his parachute. However desirable reaching an altitude of 22,000 feet would be, it requires a "cool customer" to free-fall for the approximately seventy-five seconds or so necessary to reach that altitude. It is one thing if a skilled parachutist bailing out of slow moving, easily-exited aircraft in good weather is free-falling for a certain length of time. It is quite another for a pilot to be bailing out of an aircraft traveling near the speed of sound and at high altitude — not to mention from an aircraft, the egress from which is strictly an emergency measure. Add to that the fact that the military pilot may be falling through dense clouds from his bailout altitude all the way to the ground. Cool is that pilot in such circumstances who, without the aid of an altimeter, can count close to seventy-five seconds or so during his free-fall to reach 20,000 feet before pulling his ripcord, wondering all the time whether he has counted correctly or whether miscounting will result in his emerging below the clouds with no time for a full-chute opening. And during free-fall those thoughts are blended with others concerning whether he will be conscious, whether he will be incapacitated by exposure to the cold, and where he will land.

But it was the urgent nature of the bailout at high altitude which presented the problems. If he had just experienced engine failure and felt that he did not have to exit immediately, he could have tried to reignite the engine by firing the special ignition cartridges; and, failing that, he could have stayed with the aircraft keeping a watch on the altimeter until it descended to a less hostile environment. Whatever the problem, which will probably remain forever unknown, he had to exit the aircraft at an altitude where he had to contend with all the problems accompanying that altitude. Exiting the aircraft by firing the ejection seat, regardless of its attitude, would be no problem. Of the problems described earlier — opening parachute shock, severe debilitating cold, and the need for oxygen — all of which can be disorienting and weakening — a fourth factor was at work: in Tom's situation, his location over the coastal navigational beacon and his early exit from the aircraft meant that he would be carried farther out to sea during his long descent by the generally westerly winds that predominated. Tom had known, as all squadron pilots had known, that he was required to wear the cumbersome "poopy suit" during a good part of the year when flying over ocean areas — or there was a possibility of flying over ocean areas — when the water temperature would be below sixty degrees Fahrenheit. On the day of his flight in early May, the ocean temperature was fifty-six degrees; and Tom decided that an exposure suit was not necessary, a decision no doubt partially made to avoid the time and nuisance of getting into the suit. Although the situation and his decision hardly seem worth arguing, the "poopy suit" would have provided additional warmth during the descent and, of course, in the water. Many pilots disliked putting on exposure suits because they simply felt that the chance that they would have to bail out or land in the water was so remote that going through the effort to put the suit on was a waste of time. Although wearing the suits was mandatory whenever the ocean temperature was below sixty degrees Fahrenheit and everyone seemed to obey, there was one Marine pilot in Tom's Navy squadron who told his squadron mates (but not his CO naturally) that he would turn in his wings before he would put on his "poopy suit." Not too much after his bold statement, the Marine pilot experienced a flameout (jet engine failure); and while his flight was not over the ocean and he rode the aircraft down to a rough landing on uneven ground alongside a runway that he just could not reach, squadron pilots wondered whether and how that would affect his future attitude towards wearing an exposure suit.

But in Tom's high-altitude bailout, how his decision not to wear an exposure suit when the water temperature was so close to sixty degrees affected the outcome is not known; and under the circumstances, it was not a matter anyone would discuss, given the tragic results, even though it could have been an important factor.

At the time of LT Trundel's tragic experience, most of his squadron's Cougar pilots were not aware of any Navy pilot's having bailed out at 35,000 to 40,000 feet and surviving. They did know, for example, that you could bail out right through the Plexiglas canopy in an emergency and expect generally good results; that is, that you could survive. Unknown to them though were the specifics from which that knowledge came: the bailout incident seven months earlier in which the pilot, Blue Angel Commanding Officer LCDR Arthur Ray Hawkins, USN, (later Captain Hawkins), flying an F9F-6 Cougar, did survive. Not only did Hawkins survive, but his experience also included bailing out at supersonic speed and ejecting right through the canopy. Rarely are two bailout experiences the same, because details, conditions, and locations often vary in significant ways. A few quotes from the Fall 1992 issue of the Naval Aviation Museum's *Foundation Magazine* provide some interesting details of what can only be described as frightening, harrowing conditions. But it is to be emphasized that Hawkins', like most of the incidents described exemplify Naval Aviators' dealing with and overcoming challenging, often life-threatening situations.

"Back then, I was the leader of the Blue Angels, and I had the team on a cross-country flight. All six members (having picked up new F9F-6 Cougars at the Grumman Factory in Bethpage, Long Island) were flying a comfortable line abreast when my plane went out of control. It dived into an enormous outside loop.

"I was vertical; then, I was upside down, hanging by my safety belt and beginning to red out as the centrifugal force whirled blood into my brain at tremendous pressure. That would put the fear of God into any man, especially when you know that you're traveling faster than the speed of sound. If I was going to bail out, I'd have to go now, before I lost consciousness — now or never.

"But the slipstream outside my canopy was supersonic; plunging into those granite-hard shock waves might conceivably smash the life out of me. As of this time, in all the history of Naval Aviation, no one had pierced the sonic wall with his unarmed body — bareheader so to speak. At least, no one had survived to tell about it.

After refueling in Smyna, Tennessee, "we weren't pushing our engines, but another hour would put us in Corpus Christi. Or so I thought. I had no premonition of danger. * * * I turned my head left and right to see the other planes. When I looked forward again, my nose had dropped a little below the horizon. * * * But this time, there was no response to stick movement: the nose stayed down, and it went farther down. * * * The airplane was getting away from me, diving steeper and steeper and steeper, building up speed every moment.

"Centrifugal force was pulling me out of my seat, and I realized that I was in the first arc of an outside loop. What had gone wrong? I had only seconds left to figure this out. * * * my speed was reaching dangerous levels. Machmeter and airspeed needles were wheeling around together.

"Another glance at the instrument panel showed that I was already past the speed of sound — and the needles were still winding up. Dive angle fifty degrees — vertical. The loop's centrifugal force had pinned me up against the canopy, and a crimson haze began to cloud my vision. It was the first stage of a red-out. There was one last hope: keep pressure back on the stick, full nose up trim, and switch on the flying tail (a new design which, instead of having just the elevators move, would move the entire horizontal stabilizer). * * * with that, the airplane tucked under, and I was upside down hanging in my harness. Vision was going, and consciousness would go next. There was just time enough left to jettison the plastic canopy and fire the explosive charge that would cannon me into space, seat and all. I knew the bailout drill by heart: Depress the pre-ejection lever; it blows off the canopy and arms the explosive shell behind your seat. Draw your feet back into your stirrups. Reach overhead; grasp two handles and pull the protective curtain over your face. The last inch of pull will trigger a firing pin. Boom! Out you go. But I couldn't get the sequence started. Hung up in the canopy as I was, my reach wasn't long enough to shove down the pre-ejection lever. * * * One last chance.

"Alongside my head was an emergency handle to be used only in desperate cases. It would arm the ejection seat, but it wouldn't blow off the canopy. I'd have to fire myself through the thick plastic glass. * * * I pulled down the face curtain. When the ejection charge fired, I was four or five inches off the seat. It came up and hit me like a pile driver. Too stunned to feel anything, I went through the canopy, a limp bundle traveling faster than sound. When the momentary blackout passed, I found myself clawing for my ripcord in a groggy attempt to open my parachute. Then, I thought, 'How stupid! Wait until you slow down. A chute opened at this speed would be torn to shreds.' * * * Then I realized that I was bareheaded. The wind had torn away my face curtain, helmet, and oxygen mask.

"No oxygen, altitude still above thirty thousand feet — I'd gasp my life out if I opened the parachute and dangled up here. I decided to free-fall two or three miles to get into breathing air as soon as possible. * * * Two or three miles? Why, in about four seconds, the lack of oxygen was graying me out. If I blacked out entirely, I knew I might never wake up in time to pop the chute. At an altitude later estimated as 29,000 feet, I opened my safety belt (releasing the seat) and pulled the ripcord. When the chute blossomed, it jerked the living fool out of me. The shock was so great that I thought the canopy had torn (the chute's canopy), but looking up, I saw it was intact. * * * The ground below was so far away that it didn't seem to be coming up at all. * * * Then, I couldn't see the ground or the parachute or anything;

Emergency Bailouts 141

my vision faded away. I seemed to be suspended in a gray fog. I needed oxygen. * * * I could think after a fashion and hear and feel — I remember feeling the intense cold. But I couldn't see. Then, finally, the blackout. I came back to gray, sank into blackness once more, again regained gray consciousness. The blackouts scared me. If I could only hang on until I got down where there was oxygen pressure. * * * I remembered a lesson from Navy flight training about grunt breathing: 'If you lose your oxygen mask in a high-altitude bailout, we were told, take deep breaths, close your mouth, and grunt hard. That will put pressure on the air in your lungs and force oxygen into your blood stream.'"

Needless to say LCDR Ray Hawkins did survive, landing in a cotton field, being picked up shortly and taken to the Naval Air Station Memphis, Tennessee. He reported that "For a man who had been through a supersonic bailout — an unheard of thing — I was in good shape. For instance, I might have frozen to death, floating so long in subzero temperatures at high altitudes. Luckily the slipstream hadn't torn off my shoes or gloves, and I was wearing my uniform under my flight suit. Only my ears were frostbitten"

LT Greg Burns later learned that, about a year or two after Tom Trundel's unsuccessful bail out, a Navy pilot had made a successful high-altitude bailout from an F9F-6 Cougar under conditions very similar to those of Trundel's. And much, much later, Greg would learn that there would be an ejection seat system which would address all the hazards of high altitude bail outs by automatically eliminating the necessity of manually unplugging from the main oxygen system and connecting to the emergency oxygen bottle on the parachute harness. It would automatically control the time of the pilot's separation from his seat and the altitude at which the parachute would open, all of the features being set in motion or controlled automatically, by altitude or barometric pressure-sensitive devices.

Despite all the thinking about Tom that had occupied him during the night and the lack of any real sleep he could get in the lower birth, Greg did not feel tired and actually felt as if he had no right to be tired. After all, it was the family who were really the ones under the strain. So, with the train's light jolt as it came to a stop in Pittsburgh, LT Greg Burns knew his official duties were to begin. He did not look foreword to the three days of funeral services, common in that area at the time, conducted by one organization after another, then concluded by a funeral mass and burial. But Tom had been his good friend, and he would do what he could to help, knowing his part was easy compared to what the family would go through during this time and for some undetermined future time. He, like the squadron pilots who rallied around Tom's wife at his death, and who would continue to be supportive in many ways, was ready to help. Death in Naval Aviation was not uncommon; and though each death hit hard, the pilots were prepared for it and knew their choice was to continue on. But in LT Tom Trundel's case, Greg was not prepared when Tom's five year old son approached him and said, I don't like God. I'm going to Heaven and tie Him up for taking my Daddy away."

Low-Altitude Bailout: Not a Piece of Cake

While high altitude bailouts from high speed jet aircraft carry their own special set of hazards, it should be further emphasized that all bailouts from single or twin seat military aircraft are serious, some perhaps being less so than others; but whether the aircraft is jet-powered or piston-powered, all are challenging, since getting out of the close-fitting cockpits of fighter and attack aircraft and avoiding a collision with the tail surfaces is difficult under the best of circumstances, circumstances which, in a military aircraft, always occur in emergencies. These difficulties are the reason ejection seats were designed and incorporated. Two examples, one jet and one piston powered, first the piston powered.

Ensign Sam Bridges was a young pilot recently assigned to an attack squadron flying AD Skyraiders. He was reserved but cheerful and relaxed, being careful to do everything expected of a pilot new to the squadron. As such, he was pleasant and agreeable, quickly becoming well-liked. As described in other sections of this book, the AD Skyraider had a powerful engine allowing it to carry its own weight in ordnance. Called an "attack aircraft" because of its diverse capabilities, it had twenty millimeter cannons, not fifty caliber machine guns, and twelve bomb racks which could accommodate bombs, rockets, and napalm tanks. Those racks could also carry external fuel tanks which would extend its range if so desired. Naturally, ongoing training required practice in all of these modes, gunnery flights having the easiest access to the practice area. It was one of the gunnery flights that had Sam Bridges flying his Skyraider on a bright clear day. He was the fourth man in the second four-plane division returning from gunnery runs against a towed sleeve or target off the coast of Norfolk, Virginia. Their twenty millimeter ammunition expended, the group had letdown from their gunnery altitude and were returning to the Naval Air Station, Norfolk, Virginia, at an altitude of two thousand feet. Everything seemed normal as it had on countless flights by attack pilots over many years. But today, when the returning divisions were about a mile off the coast, Sam, yelling "Fire!" over the radio, started to bail out. Such a sudden and unexpected shout startled his division mates, who then saw Sam dive over the side of the cockpit and start to fall toward the sea below. Everyone knew instinctively that such a rapid bailout was caused by more than just smoke in the cockpit. Smoke in the cockpit was not nearly so rare as fire and could come from a variety of sources, which were usually not serious. Sam's urgent bailout had to have had imperious reasons.

Strangely, after bailing out Sam did not open his chute but kept falling and falling. Since Sam was not very high when he bailed out, the pilots in the other aircraft expected to see his chute open immediately after he cleared the aircraft; and not seeing the chute open, they wondered whether there was a problem with the parachute. As Sam, falling rapidly, got closer and closer to the water, one excited pilot in the group yelled, "Pull it! Pull it!" over the radio, as if Sam were still plugged into his radio and could hear the command to pull his parachute's ripcord. But perhaps Sam did hear somehow, because his chute opened almost as if on command; needless to say, it opened just in time because, Sam was in the water a second or two later. No one knew at the time why Sam did not pull his ripcord sooner because his delay almost caused a disaster, a disaster which, in view of his resultant condition and long-persisting pain and discomfort, he might have preferred.

It was only later, after Sam had been plucked from the sea and taken to the hospital, that squadron pilots learned why Sam had bailed out so quickly yet waited so long to pull his parachute's ripcord. As to the quick bailout, there were actually visible flames coming into the cockpit licking his legs, most likely from a ruptured fuel line in the engine. Such an occurrence had been unheard of; fires don't just break out, particularly those that extend into the cockpit. No squadron pilot had ever heard of such a thing then — or for all the time they flew afterwards. Usually, that kind of fire would have to have come from a physically damaged aircraft; and while this was a gunnery exercise, his being hit by a projectile from another aircraft was ruled out by the kind of safety procedures followed. And as to the delay in

Emergency Bailouts

Sam's pulling the ripcord of his chute, the pilots still in the formation felt he might have been injured in bailing out; even the possibility that he had been temporarily knocked unconscious seemed reasonable. Although at the moment they could not be sure, the other pilots thought that he could have been injured by being struck by the horizontal stabilizer, the horizontal portion of the tail surfaces when he went over the side. If so, it would have struck him somewhere like a huge club. Later, they were to learn that it had hit him midway between the knee and the ankle, breaking his leg in several places, causing compound fractures and maybe even some shattered bones.

The pain from the break was severe; but before bailout, that same leg had been quite badly burned from the cockpit flames which had necessitated the bailout. Together, they produced such pain — even shock — as to keep him from focusing on yet a third, and still greater danger — that of neglecting to pull the ripcord and falling two thousand feet into the water. Since we cannot really experience another person's pain, we can only assume that he did rather well to be able to open the chute in the short time he had between the sudden and sharp pain and shock of severely breaking his badly burned leg and falling into the sea. And what additional pain was heaped on him when the salt water licked at his wounds?

Sam's problems were not over when he started to receive treatment in the Portsmouth Naval Hospital near Norfolk. On visiting the normally cheerful and friendly Sam, his buddies saw a very uncomfortable young man still in considerable pain, hardly aware of his visitors, even after receiving doses of Darvon, doses obviously not strong enough to ease his pain. The doctors felt that they could not set Sam's broken leg until the burns had healed sufficiently, so poor Sam lay in a prolonged state of agony until a certain amount of burn healing had taken place. Since his leg would have to be re-broken after his burns had healed sufficiently and the risk of infection had been reduced, Sam's recuperation would take some time and would not be known by squadron pilots, because they were to ship out to the Mediterranean Area in the next few days. They were encouraged by knowing that Sam would get the best medical care available — and that his young body would be a plus factor in his healing. In the meantime, his squadron mates were sympathetic and encouraging, yet they wondered whether his enthusiasm for Naval Aviation would be dampened — all the more so, when they knew that Sam was the only Naval Aviator they had known — though surely there must have been others — who became seasick seated in his aircraft while waiting his turn to take off from the aircraft carrier when there was even the slightest pitching or rolling of the flight deck.

In retrospect, squadron pilots could not really know the pain of the burns caused by Sam's cockpit fire; nor could they know the additional pain caused by his striking the horizontal stabilizer with its leg-shattering impact; nor could they know the torture of salt ocean water licking his burns and severely broken leg. They could only imagine his agony, wondering how disoriented he must have become and how, despite everything, he knew his situation well enough to act in time to save his life. Weighed against the physical injuries he experienced, and the certainty that the AD would surely be lost anyway, Sam's actions were the best of a bad situation and deserving of everyone's admiration. To put it mildly, his situation was life-threatening, and he responded with the "right stuff," as do most Navy pilots faced with challenging situations.

As is not uncommon in the Navy, no one ever seemed to hear about Sam Bridges again. It was not that the pilots were uncaring or disinterested, because many had asked about him but no one had any news. And, since the squadron shipped out to sea so soon after Sam's bailout creating an interval of several months, and on return, since many pilots were assigned new duty stations, it was difficult to learn about him. Further, it had been difficult to learn about Sam while the squadron was at sea, and many never did see him again. In the meantime, Sam was undergoing surgery to break and reset his broken leg, after which he was transferred to a rehabilitation center for therapy. Those who knew Sam, however, had no reason to think that, in due time, he would not resume his flying duties elsewhere in the Navy.

It was not uncommon not to see other pilots again once they or you were detached and reassigned to other duty. That was just the way it was then before cell phones and E-Mail. Tours of duty were usually two years or more in duration; so even in a twenty year career, you had only nine more chances of pulling duty with others you knew — and many things can happen in the meantime.

* * * *

Sam Bridges' bailout was necessitated by something over which he had no control and his injuries were caused in a bailout from a relatively slow moving aircraft powered by a piston engine and a huge propeller. Another bailout was necessitated by an incident that was likewise beyond the pilot's control. In this incident, there is a factor, which, when examined, some would claim included flight technique. There is no intent here to be critical of the pilot — how could anyone, given the circumstances? The F6U-1 Pirate, built by the Vought (formerly Chance-Vought) Company did not enjoy a good life. Rumors among pilots were that it was underpowered, rumors that were later described in publications such as U.S. Naval Fighters by Lloyd S. Jones. Its Westinghouse J32-WE-30 engine, producing just 3200 pounds of thrust, was not enough. Even the addition of an afterburner was not enough because of the additional weight of its installation. As a result, only a few were produced, and they were not used in fleet operations. So, when LT James Finch was assigned a test flight out of NAAS Mustin Field in Philadelphia, he knew its power; and he also knew that the F6U-1 Pirate had had some modifications made by aviation mechanics at the Naval Air Material Center there. When the aircraft was designated as ready for the test flight, Finch, an experienced pilot, taxied out to Runway 15 (heading 150 degrees) for preflight checks and a review of the Takeoff Check List. There were no taxi strips at NAAS Mustin Field, so the pilot had to taxi down the active runway to its end, where a large, circular 400 foot turnaround space enabled the pilot to pull to the side of the runway and perform his preflight checks. The lack of taxi strips necessitated that the turnaround areas be big enough so that the aircraft had room to turn around, if the runway itself was too narrow, and that they provide a space clear of the runway where the pilot could perform his preflight checks and where the aircraft's location not interfere with landing traffic. So, although Runway 15 was listed as 4000 feet long, the turnaround areas at each end of the runway added 400 feet to each end of the runway, allowing the pilot to use 4800 feet for takeoff, provided there were no obstacles to clear at the end of the runway which would require him to get airborne well before he reached the end. Runway 15, extending from the northwest to the southeast, headed toward the Delaware River, not much after the river's flow changed from a generally southerly direction to a generally westerly direction. Given what was to follow, the change in the river's direction of flow was fortuitous. Taking off from Runway 15 and maintaining this direction for awhile, a normal takeoff pattern to allow the aircraft's airspeed to build up while the aircraft gained altitude, the straight ahead direction would have the pilot crossing the Delaware River at an angle of forty-five degrees, in effect making the river crossing much longer — broad at this point anyway, the diagonal crossing made it much wider.

His check list now completed, Finch received takeoff clearance from the tower and taxied into position for takeoff. He revved his jet engine up to 100 percent of RPMs, maximum power, released his brakes, and started his takeoff roll down all 4800 feet of runway. But on this day of light, steady head winds right down the runway and an aircraft void of ordnance, thus having relatively low gross weight, and an underpowered engine, 4800 feet of runway would not be enough. What happened next was quite unusual. Finch, an experienced pilot, no doubt, knew about takeoff technique. Technique could conspire to increase drag thus tending to hinder acceleration down the runway. If the pilot raised his nose wheel off the ground too soon to increase the angle of attack of the wing for takeoff before sufficient takeoff speed had been reached, the resultant drag or air resistance could keep the aircraft from getting airborne in the requisite distance — despite full power. Keeping the nose wheel on the ground until sufficient takeoff speed has been reached is essential

Whatever technique Finch employed with his use of 100 percent of power, and there is nothing to indicate he used faulty technique, at the end of the 4800 feet of runway he could not clear a height of about sixteen inches, the height of a concrete curb and a railway tie of about four inches, both of which were at the edge of the river and about four feet above the river's level. Finch may have been barely airborne or just lifting off the runway, when on striking the curb and/or railway tie, the nose wheel, one main wheel and strut, the other main wheel, and one tip tank snapped off instantly. But strangely, the aircraft seemed to have bounced into the air and was now airborne, flying barely above both the stalling speed and the river's surface, so close that the river's surface was disturbed by the blast of the jet engine's exhaust. Maintaining the runway heading — any turn, however slight, would require more lift and thus be dangerous — he gradually built up speed to enable the aircraft to climb so that, by the time he was well across the river, he had achieved sufficient airspeed to clear obstacles on the other side of the river. One can only imagine Finch's thoughts during this whole process. However, as soon as Finch determined that the aircraft might be flyable, his immediate concern was a bit lessened. But whatever else he would have to do, the first thing would be to take inventory to see whether there were problems which could affect the degree to which the aircraft could remain in flight and what equipment was still operable. He determined that he still had electrical power and a working radio, so he contacted the control tower at Mustin and asked them for a fly-by so they could see and describe his external condition as he made a low pass. He had, of course, known that he had struck the curb hard enough to be sure he had done some damage, and he could see that he had lost a tip tank. But what about his landing gear? He had assumed that there was damage, perhaps considerable; but he needed to know how much. Advised that he had lost his wheels plus the nose wheel strut and right wheel strut, he knew his condition was serious. But he was further advised that, despite the loss of his left wheel, the left wheel strut remained in an apparently locked "down" position — not exactly good news. The good news, however, and not appreciated at the time, was that the aircraft was very rugged, rugged enough so far to keep him alive.

But what now? A landing under the circumstances was impossible. The fixed left wheel strut made a wheels-up landing out of the question. The only thing left was to eject from the aircraft. Once sufficiently calmed down and after his consultation with the control tower people, Finch announced that he would eject, a decision which, under the circumstances, no one could argue with. A suitable place, one with minimum risk to inhabitants, was decided upon, and Finch proceeded to that area. Since the ejection seat was powered by a cartridge or shell containing explosive powder, it was not dependent on the aircraft's electrical or hydraulic systems. It was just a matter of flying to the designated area, setting the pre-ejection lever to get rid of the canopy, positioning his legs back in the stirrups close in to the edge of his seat, and then reaching up for the face curtain handles and pulling them and the attached curtain down over his face and shoulders until the handles reached his waist, at which point the cartridge would fire and send him and his seat well clear of the aircraft. Having accepted the idea that he must bail out and being well prepared for the procedure he had to follow, Finch expected it to be routine — and it certainly should have been. After all, ejection seats always worked; that is, in all cases but his.

Repeatedly, Finch pulled the handles of the face curtain down to his waist, expecting each time that he would be hurled clear of the aircraft. It is easy to imagine the frustration and the uncertainty occasioned by each pull, and then his wondering whether the shell might fire somewhat accidentally when he was sure it would not, thinking also that, under the circumstances, he might be careless and not have his legs protected for the ejection. After many frustrating attempts to eject, Finch determined that his exit must be via a different route; and after due consideration, he decided to roll into an inverted position and hope that he could just drop out of the cockpit. Not sure of what problems that might bring about, particularly whether he might be struck and severely injured by the tail surfaces, he nevertheless decided to chance it and rolled onto his back. With the kind of luck he had so far experienced, he could reasonably have expected to be injured in the bailout and that maybe the parachute would not function. On his back, he released his seat belt and dropped free of the aircraft uninjured; he then descended safely to the ground. No injuries, no parachute problems, only relief that, having endured all of the problems he had faced, he could walk safely away from some very unpleasant and trying incidents. He had handled his emergency, he had contended with a faulty ejection seat system, he had made an unexpectedly good bail out, and he had persevered to make the best out of a very challenging situation.

Any pilot having just experienced a serious incident does some serious thinking, often trying to find reasons for things, sometimes blaming himself, and wondering whether the entire thing could have been avoided. With the underpowered Pirate, he would blame the aircraft to some extent for his takeoff problems. Surely, he expected to be sufficiently airborne at the end of 4800 feet of runway. What had he done wrong? Had his takeoff problems stemmed from his takeoff technique? Underlying all, was his trying to resolve how his striking the curb at the end of 4800 feet could have affected his ejection seat system. He was about to learn that there was no connection.

Not too much later, but after he had been safely returned to base and examined for injuries, the mystery of the faulty ejection system was cleared up quickly. He was informed by the mechanics who had worked on the airplane, that there was no way Finch could have known that the ejection seat system had been disarmed. Unbelievable! It had been disabled so that there could be no accidental firing of the seat while one of the technicians was working on the aircraft's cockpit while the plane was inside the hangar. There had been rumors of a very rare occurrence, perhaps more, when the ejection seat had been accidentally fired while someone was working on the airplane in the hangar, so the technician's safety procedure had some rationale — that is, as far as the mechanic's safety was concerned. Whatever the reason — oversight, carelessness, etcetera — the ejection seat firing system had not been returned to a ready-for-flight condition. After the incident, the fact that the disabled ejection seat firing system problem did not receive much attention at the local level did not preclude reports of the incident from being widely circulated so that a recurrence would not be very likely. As for Finch or any other jet pilot, it was not practical to inspect the ejection seat system before a flight. It was not required on either the preflight check list or the takeoff check list; it would not have been easy to check because in many cases it was not easily sighted. Like the confidence that pilots have that parachutes will always work properly, confidence in ejection seat operation had to be relied upon as well. It was bad enough that Finch had had the harrowing takeoff experience, but then to have to exit a jet aircraft in an unorthodox and dangerous manner was discouraging to say the least. Although the Pirate was a bit underpowered, nothing but positive things could be said about the ruggedness of the airplane that did not just break up or disintegrate when it failed to clear the cement curb and railway tie. In spite of everything and in fairness to LT Finch, a runway of 4000 feet to 4800 feet is hardly suitable for the demands of daily jet aircraft operations, even if under some circumstances where the wind force is strong and right down the runway, a much shorter takeoff or landing distance is suitable.

For as long as he lives, LT Finch will hold the Delaware River to be special, a place of special reverence in his memories: for if he had not had the broad stretch of the Delaware River in front of him, he would never have survived. How fortunate he was that the Delaware River turned west at the east end of Mustin Field and about one mile before Runway 15 intersected it; likewise, he was fortunate to have crossed the Delaware at a forty-five degree angle, extending the obstacle-free space. Other pilots, and no doubt Finch as well, shuddered to think what the outcome might have been had Finch had to take off in the opposite direction that day where much greater and higher obstacles prevailed.

The reader is surely wondering why the pilot bailing out of the AD Skyraider, a much slower plane, was injured, while LT Finch with a disabled ejection seat system just rolled over on his back and dropped out. The most likely explanation is that LT Finch had slowed his aircraft down as much as possible and was able to exit at his choosing in a less than emergency mode. Ensign Bridges' exit in contrast was in a more urgent mode where he did not have time to control factors which might have lessened risk; his exit had to be fast and the flames constituted an emergency requiring immediate action.

* * * *

Relatively low altitude bailouts like those described here seem contradictory, considering that the pilot bailing out of a jet aircraft with a disabled ejection system was not injured, while the pilot bailing out of a piston powered aircraft was severely injured. Are there any in-betweens? In what seems to be a one-of-a-kind incident, a young AD Skyraider pilot had a bailout incident that was very strange, even bizarre. No judgment is made here, because, after all, the reason for his bailout was engine failure, something over which he had no control. But, forced to bail out at a low altitude, and for whatever reason — perhaps fearing he was too low — he pulled his parachute's ripcord too soon. The blossoming parachute became hooked on the Skyraider's tail. The aircraft, now pilotless and towing around a snared parachute with the pilot still in its harness, continued its slow descent to a reasonably gentle touch down — the pilot still in tow! Recollections of this incident, no doubt verifiable by Naval Aviation News' Grandpa Pettibone, are that the pilot survived. Whatever the reader may infer from this incident — poor technique, a chance entanglement, or just poor luck — the pilot did not cause the engine failure, which was the reason for the bailout; and he had just a second or so to make a decision and act on it.

There used to be a saying that "any landing that you can walk away from is a good landing." The question is, "Does this one qualify?"

Bailouts on the Deck and Underwater: It's Hairy

Some reference to "Zero Lanyard" or "Zero, Zero" parachutes was made at the beginning of this section. The author has personal knowledge of five or six such bailouts. In most of these, and provided there were no other complicating factors, such as inverted flight close to the ocean surface, egress was accomplished satisfactorily, if sometimes far from comfortably. But even in those cases where a fatality occurred, the ejection seats and parachutes functioned as designed.

Any time that circumstances require the pilot to eject while the aircraft is on the ground without flying speed is quite tense. There is a decision that must be made in a second or two. A pilot taking off and experiencing engine or other problems quickly becomes faced with the reality that these problems are going to keep him from getting airborne safely. There is often a moment's hesitation for him to face and realize his problem; that is, there is a moment to wonder whether ejecting is the right course of action. Then there is a slight concern about whether his parachute will perform as well as they always have, a concern that might make him wonder about just staying with the aircraft as it runs off the end of the runway or aircraft carrier deck and face what may be waiting there. In almost any case imaginable, other than shutting down just at the start of the takeoff roll, staying with the aircraft is extremely risky. Yet, if any situation can be worse than bailing out while still taking off from a runway, it is bailing out while the pilot is being catapulted from an aircraft carrier at midnight in mid-ocean. Such an incident happened to LCDR James "Max" Qualls and his Radar Intercept Officer (RIO) LT Mike "Mouth" Mealy on one of those dark, moonless nights know as "dark-ass nights" by carrier pilots flying out over the ocean, at a time when all is blackness — no horizon, no ups, no downs, no visual references — leaving only the instruments in the cockpit to provide information.

LCDR Jim Qualls was a Naval Aviator with much flying experience. Besides extensive F-14 Tomcat experience, he was a Navy flight instructor and has flight time as a C-9 Skytrain II Aircraft Commander. His twenty-six years of flying include duty with United Airlines from 1996 to the present time. He was a part of the mission "Operation Provide Comfort" to assist the Kurdish refugees after "Operation Desert Storm." Qualls had a zest for flying which includes many memorable incidents: a five week power projection course at Topgun, which allowed him to achieve a speed of 720 KIAS (air speed) or 1.1 IMN (indicated mach number) in his F-14 Tomcat and blow by F-18 Hornets as if they were standing still, a feat which "gave him the pleasure of watching the jaws hit the floor in the TACTS debrief as they watched the playback; on another occasion, he became retchingly ill with food poisoning while flying the Tomcat, vomiting into his gloves and elsewhere — yet still having the necessity to fly back aboard the aircraft carrier on one of those "dark-ass" nights — his only choice. One of his expressions seems a clue to his approach to flying: "Yes, flying the F-14 was about the most fun you could have with your clothes on." And concerning combat, "Yes, speed is and always will be LIFE!!" What follows is a narrative based on communications with Jim Qualls, including quite a few E-Mails and some telephone calls. At some point in these communications, it became apparent that Jim's recollection and his description, written in his hand, would provide the best way to present his story. Jim Qualls states, " I remember every detail of that ejection sequence even today, some twelve years later."

"How many sea stories have you heard that started like this: 'It was a very dark night… ' I know; I've had plenty. And that's because dark nights make for such good sea stories. Mine are no different. It turns out it was not the night to swap airplanes with my nugget wingman, but that is how this episode began. The jet had a history of losing the VDI (primary attitude indicator) on the catapult stroke. Wow! 'Sure, I'll take that one.' Like I needed another item to help the pucker factor!

"We were operating off the coast of Florida on a cool night in December of 1990. It was just about midnight and we were set for the launch. I taxied the big Tomcat over to catapult #4 on the *Forrestal* (*USS Forrestal* CVA-59). This night was no different from most others. However, as most intrepid aviators who fly from ships can attest to, no two are the same. They all have their own unique set of events that make each one so memorable.

"Sitting on Cat #4 on the *Forrestal* at night reminded me of the first time I anticipated running toward a brand new slip and slide. Oh sure, you know it's wet but what if you hit a dry spot? That's gonna hurt! Anyway, the light goes vertical from the catapult officer and I stroke the burners of the mighty Tom and we are now being held back by the ship with a bunch of steam waiting to hurl us into the night sky! I know that sounds a bit dramatic, but this is a dramatic moment! Let's not forget, this is the only part of the night's events that I have absolutely no control over! Sure enough, WHAM! The catapult fires and fails to take us with it! Apparently, the young airman whose job it was to insert the launch bar properly into the catapult shuttle forgot about the *'properly'* part! Well, the F-14 is a powerful jet, but not powerful enough to get airborne in 300 feet without the help of some highly compressed steam!

"Now, the Air Boss, one CDR Frank *'Krank'* Kramer, is watching from the tower and fittingly and quite loudly shouts *'Eject, Eject, Eject!'* over the UHF radio which we have tuned in the cockpit. But, let's go back for a moment to when the holdback fitting 'broke,' the catapult leapt forward (without us) and hurled itself into the water break, and we were left trundling down the deck in zone five afterburner with not nearly enough 'runway' to get this beast into the air. If this happens to you, you will know it immediately. There is nothing quite like the feeling of a properly executed cat shot on any ship (and I have flown from about 11 different decks). You know if you're going flying! Well, this night I knew we were not. At least not in an airplane! We did go flying as the ejection seats hurled our pink bodies into the dark night!

"Every detail of that ejection sequence seems to be indelibly burned into my memory. Let's just say that it happened very quickly, but it made a lasting impression on me! The canopy being jettisoned was very loud and then I was in total darkness! One moment I was in the F-14 with hands on the stick throttles (now at idle) and feet firmly on the toe brakes, and the next moment I was in relative silence with a tumbling sensation. As I got my bearings back, I realized we had been ejected. Mike Mealy, (callsign *'Mouth')* was my RIO that night. He saved our butts! After we heard the dreaded 'E' word yelled at us in triplicate over tower frequency, *'Mouth'* uttered a very profound and decisive expletive, *'O...* ' (you fill in the word).

"These two words are the international signal for: *something bad is about to happen unless you do something else very quickly to get yourself out of this bad situation*. So he did; he pulled his lower ejection handle. The rest is history. I made a very firm two point landing (on the carrier deck) which my ankles and back have yet to recover from some thirteen years later. Now, because my chute was still inflated, my back, shoulder and ass were enjoying the sensation of being *'sanded'* by flight deck non-skid. I guess I just hit a dry spot on the ultimate slip-and-slide! Thank God, some nice young man on the flight deck jumped on my chute and deflated it before I had run out of skin on my right shoulder. At any rate, we landed; the jet came to a complete stop after doing a right 90 degree turn.

"Moral of the story? Ain't one; just be ready to go. The next time you climb into your cockpit and arm the seat, remember that you are arming it for a reason. It's like a cop who carries a gun fully loaded with a round in the chamber. They put ejection seats in airplanes for a reason. Whatever you do, *Don't ride one to impact!"*

And what about LCDR Mike *"mouth"* Mealy, whom he credits with acting in a timely manner to "save our butts!." Qualls reports further: "My RIO, who weighed at that time 150 lbs. soaking wet went about 50 feet higher in the air than I did. His chute caught on an antenna on the signal bridge and he walked away uninjured. He got about a full swing in the chute and was airborne for about 10 seconds. At the time, I weighed about 190 lbs. I got half a swing and was airborne for about 6 seconds. I landed amidships aft of catapults 1 & 2 just about abeam the island. Our time to react (before ejecting) was just about immediate to keep the jet on deck. I immediately applied full brakes and retarded the throttles to idle which kept the airplane on the deck."

There was no way that the Air Boss Kramer or Qualls and Mealy could be sure the aircraft would not go over the bow and into the ocean. Such an outcome would have carried unthinkably great risk. So, decisions were made in an instant, correct decisions which precluded losses which occurred in other incidents where loss of life was the result. It is interesting to recall Qualls' statement in another situation at TOPGUN where he stated, "Yes, speed is, and always will be LIFE!!" As to his other statement, "Yes, flying the F-14 was about the most fun you could have with your clothes on," that is left to one's imagination.

* * * *

The question naturally arises as to whether, if an ejection seat system and parachute which are designed to function in the fluid of air, where they can operate to free the pilot from the aircraft and can capture and release air to lower the pilot slowly and safely to earth, have any effectiveness when the aircraft is submerged in the fluid of the ocean. How then by what means, can the system come to the rescue of a pilot so submerged? Contrary as that may seem, such an occurrence has not been a solitary incident. There have been a number, though not a host, of such experiences. What follows is a general description of what takes place under water as quoted in an article on the web site < www.ejectionsite.com/eunderh2o.htm > That is followed by a first-person article of an actual under water experience by CDR Russ Pearson USN, (Retired,) which appeared in The Hook magazine, a publication of the "Tail Hook Association." A large part of that article is presented, naturally by obtaining the association's permission.

From the website: "Probably the rarest form of ejection, ejecting while submerged. As odd as it may sound, it is feasible and has been done successfully. Once submerged, it is virtually impossible to open the aircraft canopy against the pressure differential between the water and air in the cockpit. Once the cockpit is full of water, it might be possible to slowly push the canopy open and exit the craft, but the amount of time necessary for the cockpit to fill with water would allow the plane to sink below the depth the pilot could survive. The pilot's oxygen mask is optimized for use in thin atmosphere conditions, and cannot be counted upon to provide breathable oxygen under any depth of water.

"The above factors mean that there is only one significant option available to a pilot once the craft becomes submerged — eject through the canopy. The same forces that prevent a pilot from opening the canopy manually would prevent the jettison charges from pushing the canopy safely out of the way of the ejection seat. The seat does not usually exit the water; it merely crashes through the canopy and then initiates seat separation. The pilot must then separate from the parachute that would usually partially deploy, and swim to the surface, not necessarily in that order."

From CDR Russ Pearson's article *Underwater Ejection*: "Shortly after midnight in the 'zero-dark-thirty' hours of 10 June 1969, I was the pilot of a single-engine, single seat A-7 Corsair II light attack aircraft that departed the flight deck of the aircraft carrier *USS Constellation* (CVA-64) and plunged into the Pacific Ocean some 60 miles off the coast of Southern California.

"The mishap occurred at the end of a marathon 23-hour day that culminated with the first of six scheduled night carrier landings. The event was to have marked my final night of initial carrier qualification (carqual) training as a fleet replacement pilot with VA-122 at NAS Lemoore, Calif.

"THE LANDING: The voice of Connie's final approach controller came through the headset loud and clear, 'Corsair 202 is on course, on glideslope at three quarters of a mile. Call the ball.' It was my cue to get off the instruments and fly the final few seconds of the approach visually. A light drizzle was falling from the low hanging overcast just above the landing pattern, but the visibility was good underneath and the sea state calm. The A-7 Corsair II aircraft strapped around my waist was the Navy's newest light-attack carrier jet and I was proud to be in one of the initial classes of first-tour pilots selected to fly it. 'Two-Zero-Two, Corsair, ball, fuel state 4.0,' I replied as my scan shifted outside the cockpit to the meatball of amber light beaming aft from the optical landing mirror on *Constellation's* four-acre flight deck. The seat of my pants told me I was too high, but the ball was centered on the mirror to confirm I was on glideslope. My 4,000 pounds of fuel was a comfortable reserve, ample to make it around the landing pattern a couple more times and still have enough fuel to 'bingo' to the primary divert field at NAS Miramar if I didn't get aboard.

"'Roger, Ball. Keep it coming,' the landing signal officer (LSO) acknowledged from his platform on the port side of the flight deck. The voice was not as relaxed as the LSO who had 'waved' the class every night for the past month at Lemoore.

"More than two years of flight training at five bases in four states were riding on this event. Tonight was the long-awaited 'graduation exercise' from the training environment into the fleet, the final rite of passage into the Navy's elite fraternity of tailhook carrier pilots. In a few short months, I'd be flying combat missions in Southeast Asia from an aircraft carrier in the Gulf of Tonkin.

"Scheduling such a significant event at the trailing edge of a grueling 16-hour day should have raised caution flags somewhere, but not with me. The instructor pilots had primed the class for months with sea stories about night carrier landings separating the 'men from the boys' — now it was my turn to prove I could fly with the eagles. The adrenaline was pumping.

"The non-stop day that began with a 0330 wake-up call back home in Lemoore had been a test of endurance, but long days are part of the normal routine aboard carriers at sea. Besides, we were training for combat, and 'hacking the program' was part of that training — this was the Navy, not the airlines. The squadron's mission was to pump out combat replacement pilots for NavAirPac's light attack Corsair squadrons, and pilot output was running behind schedule. The pressure was on from the top down to catch up. In the light attack community, 'death before dishonor' was the unwritten code. Begging off a flight schedule, especially with a flimsy excuse like fatigue was a sure way to be branded a 'non-hacker' for the rest of your career.

"The final half-mile to the ship was over in a matter of seconds — it happened so fast that the tricky 'burble' of turbulent air at the fantail passed practically unnoticed. But the bone jarring jolt of the 25,000 pound Corsair coming down at 650 feet-per minute to collide with the ship's steel deck didn't go unnoticed. I knew it was coming but it still got my attention. The harness straps dug deeply into my shoulders as the plane decelerated from 135 knots to a screeching halt in three seconds flat. The first night 'trap' had

Emergency Bailouts 161

lived up to its billing: it was a cross between ecstasy, and a head-on collision with a freight train.

"'Piece of cake,' I thought. 'Five more and you're on your way to the fleet.'

"The landing was on speed and on glideslope, and the tailhook had engaged the targeted No. 3 wire. All was not well, however, as the plane was drifting fast toward the port catwalk. On this, the fifth man up, third launch and eighth trap of the extended day, fatigue had finally overpowered my adrenaline. I had become so focused on flying the ball that the landing center line had momentarily dropped out of my scan. A late line-up correction had set up a right-to-left roll-out as the plane decelerated down the angled deck.

"OVER THE SIDE: The plane skirted the port deck edge like a tightrope walker on a high wire before stopping painfully close to the catwalk. I couldn't believe this was happening to me — could already hear the lineup lecture from the LSO back in the ready room debrief.

"The cockpit was jolted hard as the plane's port main landing gear dropped off the deck edge. As luck would have it, the protective steel scupper plate guarding the deck edge had been removed during the ship's recent trip to the shipyard and had not been replaced. In less than a heartbeat, the plane was precariously perched on the edge of the flight deck.

"It was hard to tell the plane's exact attitude with no visible horizon, but the fuselage was turned at least 60 degrees left-wing-down. To eject now would be suicidal — the trajectory of the ejection seat's rocket motor would send the seat skipping across the water like a flat rock on a farm pond. If the hook remained engaged with the arresting gear cable, the situation might still be salvageable.

"As the magnitude of the situation settled in, my mind suddenly shifted into slow motion. Strangely enough, there was no panic — at least not yet. My thoughts were surprisingly calm and clear as I instinctively pulled the throttle aft and 'around-the-horn' to shut down the engine. If the hook should release from the cable and the aircraft went over the side, the prospect of cold sea water combining with the Corsair's hot power plant was a recipe for an even more explosive situation. The engine was of no use now anyway.

"As the engine spooled down through 65 percent rpm, the generator dropped off the line and cut off all electrical power — as the radio and interior lights went out, total darkness immediately enveloped the cockpit. All contact with the world outside was lost. I had been alone in a crowd before, but never like this. Except for the pounding in my chest, there was only dead silence and it had a deafening sound. If this was a dream, it was a nightmare! Unfortunately, I wasn't dreaming.

"The momentary stillness was soon shattered as the aircraft lunged forward. The worst had happened — the tailhook had 'spit-out' the arresting cable. I was in deep, serious trouble and knew it. The plane tumbled off the flight deck and plunged downward some 60 feet prior to impacting the Pacific — the sensation was like falling into a black hole.

"We had learned in survival training that a ditched aircraft normally sinks at about ten feet per second, and after 100 feet, crew survival is highly unlikely. I figured I had about ten seconds if I were going to get out of this mess alive. It appeared that only a miracle could save me now. I had just run out of altitude and airspeed, and was about to run out of ideas, too.

"The ejection seat seemed the only choice, albeit a slim one. In the history of Naval Aviation, only a handful of pilots had ever attempted, much less survived, an underwater ejection. It was theoretically possible in the A-7, but no one had tested it.

"There was also the chance I might eject directly into the Connie's passing steel hull or even worse, into one of the massive propellers. The odds for survival were grim and getting worse each second.

"EJECTION: I intentionally delayed the inevitable for a split-second for the ship to pass clear. Then, like a death-row prisoner condemned to throw a switch and end his own life, I reached down between my knees for the seat's alternate ejection handle, the one we'd been trained to use when time is the most critical factor. Images of my wife Theresa waiting at home with Steve, our nine month old son, flashed through my mind. How would she react when the black Navy sedan pulled into the driveway and the skipper and chaplain came to the door. Realizing this might be my last conscious thought, I grasped the ejection handle, closed my eyes and, expecting the worst, pulled straight up… nothing happened. Time seemed to stand still.

"The delay was only a millisecond, but it seemed much longer. I had already decided that the ejection seat was not going to work and saw myself slowly sinking, to drown or be crushed to death by the depths. The Corsair's tiny cockpit seemed destined to be my coffin.

"A sudden blast of brilliant light blinded me — the seat's rocket motor had fired following a built-in sequencing delay. In an instant, I was out of the cockpit and clear of the seat, though still submerged in the cold, dark water of the Pacific.

"FOCUS ON SURVIVAL: I couldn't breathe. The water had forced the oxygen mask down around my chin and the emergency oxygen bottle in the seat pan was useless; for the first time, panic set in and I became totally disoriented — I couldn't tell up from down. It was as if I had been shot out of a high-powered cannon into a pool of jet-black ink, a far cry from the Dilbert Dunker simulator in the crystal clear water of the training tank back at the Water Survival School in Pensacola. In less than a minute, I had gone from being a cocky, self-assured carrier pilot to a desperate young 25 year-old Navy LTJG fighting for his life.

"I had to do something fast or it was all over but the memorial service. Just then, a cluster of lights flickering of the surface caught my eye. As an 80,000 ton aircraft carrier cutting through the water at 30 kts doesn't stop and turn around on a dime, the flight deck directors had tossed their water-tight flashlight wands over the side to mark my plane's location for the plane-guard destroyer and the search and rescue (SAR) helo. Though I was still under water, the lights reoriented me and I instinctively swam toward them.

"I gasped for air as my helmet broke the surface — it felt great to be alive. But that lung full of fresh sea air was accompanied by an excruciating pain as if a butcher knife had been plunged between my shoulder blades and twisted. Something was seriously wrong, but there was an even more pressing problem.

"The altitude-sensing device that automatically deploys the parachute had activated and the chute had partially opened. The canopy and its nylon shroud lines were streaming behind me, overpowering my frantic efforts to keep my head above water. I had to stay clear of those shroud lines and get rid of that chute, now.

"I grabbed for the nylon toggles that inflate the lobes on the Mk 3C survival vest, but they weren't where they should have been. Panic began to set in again and time was running out. I was fast losing the struggle to keep my head above water — it took all the strength I could muster just to say afloat. The parachute was winning and I was on the verge of being dragged under.

"My body suddenly went numb as something below the surface brushed against my feet. During the ejection through the Plexiglas canopy, my left forearm had been sliced and was bleeding profusely. The gash on my arm was even more reason to be alarmed. The survival vest contained several packets of shark repellent but I was too busy trying to keep my head above water to get to them. When the object brushed against me again, I realized that it was the plane. It had impacted the water with minimal force and was virtually intact. With its wing fuel bladders and over half the fuel cells filled only with air, Corsair 202 was floating upside down just beneath the surface, still bobbing from Connie's passing wake. I had surfaced alongside the aircraft and my legs had brushed against the tail. Hanging onto the horizontal stabilizer for support, I finally located the life vest's inflation toggles which had wrenched around to my side during the ejection. Grasping a lanyard in each hand, I pulled down and away and whoosh, the flotation lobes inflated instantly.

"But I wasn't out of harm's way yet — the parachute still streamed out like a huge sea anchor. Should it fill with water and sink, even the inflated life vest wouldn't help. I glanced around just in time to see Connie's plane-guard destroyer bearing down on me. From my water-level perspective, the 'small boy' looked anything but small, and if she didn't change course, the rescue part of the mission would be over and recovery and salvage operations would begin.

"Using techniques learned in water-survival training, I rolled over on my back and reached upward along the parachute risers until I located the koch fittings, the small metal latches that connect the harness to the parachute. I lifted up on the cover and pulled down on the latch. In an instant, the chute was gone.

"SAR HELO TO THE RESCUE: Moments later I was floating center stage in a large beam of bright, white light shining down from the ship's SAR helicopter that hovered noisily overhead. Like most jet jocks, I had never fully appreciated helicopters except when they brought the mail — they had always been high on my list of low-priority aircraft. Never again! Just now, that homely, wind-blowing, water-churning contraption looked like an angel of mercy — nothing could have been more beautiful. Fortunately, the destroyer had veered off to starboard and yielded to the helicopter.

"Minutes later, a rescue swimmer from the helicopter was in the water next to me.

"You okay, sir? he yelled over the din of the thrashing rotor blades.

"'I'm okay,' I yelled back, 'but it hurts to breathe.' I didn't tell him that it also hurts to yell.

"'Hang on sir. All we've got is a horsecollar, but it'll get you out of here,' he shouted as he guided my arms through the opening in the pear-shaped rescue sling that nestled under my armpits.

"As the hoist began lifting us slowly out of the water, my body dangled helplessly from the horsecollar like a wet dish rag. Weighted down by soaking flight gear and steel-toed flight boots, and whipped about by the helo's downdraft, the pain became unbearable. The next thing I remember was sprawling on the deck of the helo's cargo cabin, heaving salt water.

"Moments later, the helo recovered aboard the carrier and I was transported to sickbay on a stretcher. The alternate ejection handle may have expedited my exit from the cockpit, but at a painful price. Reaching down between my knees to grasp the secondary handle in an inverted, submerged cockpit had placed my upper body in a vulnerable, dangerously curved position. The brutal g-force of the seat firing had broken my back.

"THE MIRACLES CONTINUE: Three days after the mishap, the ship's senior medical officer, a newly selected Navy captain, arranged to accompany me ashore on a MedEvac flight to Balboa Naval Hospital in nearby San Diego. By coincidence, the flight was scheduled with the same crew and aboard the same helo that had rescued me earlier.

"Just prior to boarding the flight, a casualty on the flight deck created an unexpected dilemma — the MedEvac helo was configured to carry only one patient. The doctor had an instant decision to make. Needless to say, I was not happy to learn my name had been scratched from the manifest only moments before the launch.

"About an hour later, a young corpsman came running onto the ward. He was out of breath. From the look on his face, I knew something terrible had happened.

"'You're living right or somebody's looking after you, Lieutenant,' he blurted out. 'Word just came down from Air Ops that the MedEvac flight had engine problems and went down in the water about halfway to the beach. The crew got off a Mayday and another SAR helo found the wreckage right away, but there were no survivors. Not even the Doc.'

"I respectfully declined a second chance to MedEvac ashore, electing instead to ride the ship back into port a few days later. Shortly after *Constellation* moored at the carrier pier at North Island, the corpsmen carried me ashore on a stretcher to be transported by ambulance the short distance to Balboa Naval Hospital.

"Naval Aviation had turned out to be as dangerous as it was glamorous. In three short days, I had cheated death twice and, in the process, learned firsthand that the thrill of flying high-performance jet aircraft off the decks of aircraft carriers sometimes demands a hefty personal price. I now understood why guys get paid for a job most of us would gladly pay for the privilege of doing.

"Whether my survival was fate or just sheer good luck is debatable. Maybe the corpsman was right — maybe someone was looking after me. But one thing is for sure. Without the first-class water survival training all tailhookers receive as they earn their Wings of Gold, I would have been remembered by friends and family, at a 1969 memorial service rather than honored by them at a 1992 Navy retirement ceremony. Every day since 10 June 1969 has been a gift of life for which I am thankful."

That CDR Russ Pearson USN continued flying for many years until retirement tells us something about Naval Aviation and its pilots. So many are of the Pearson mold, determined to persevere in the most challenging of situations. As to the carrier landing which led to his underwater bailout, no one who knows the nature of landing fast, heavy, fuel-filled aircraft back aboard ship on dark nights can really be critical. The procedure is best described by a seasoned veteran. In the April/May 1995 issue of *Air and Space Magazine*, "CDR Chris Nutter, a veteran of 700 traps (carrier landings) and the executive officer of VFA-137, an F/A-18 squadron on the *Constellation*, tries to convey to non-pilot friends the sensations of the night landing: 'Imagine that you're in a car without headlights going 150 miles an hour down a narrow dark road toward a one-car garage illuminated by a single light bulb. If you get through the garage door, your car will stop automatically. And the garage is moving around. That's what the night landing is like." No one having had the experience of night carrier landings could disagree.

Self-Induced Bailouts: No Piece of Cake Either

Perhaps this section's title leads one to infer that inducing your own bailout should really makes things easy, if not exactly a cinch. The reader, knowing that bailouts from single-seated military aircraft are considered emergencies, should not make such an inference. No doubt, some Naval Aviation agency somewhere — China Lake, California, most likely — may have records and details on incidents in which Naval Aviators have had to bail out. But those not privy to such data are left with only their personal knowledge. Logic would indicate that no one would really want to induce his own bailout, so it is not such a stretch of the imagination to think that such bailouts are not numerous. However, two contrasting bailouts are described here — contrasting because of the different types of aircraft and the difference of personalities of the Naval Aviators involved:

The first was caused by an extremely confident, self-important young aviator's mistake in judgment. Stephen Burg had the reputation of being an exceptionally competent young Naval Aviator. A junior officer (lieutenant junior grade), he had accumulated a fair amount of flight time — and in some of the Navy's latest models. He had been known to show off a bit, as when he had a fellow aviator in the back seat of a TV-1, a version of the Air Force's P-80 Shooting Star. The two-seats in the straight wing jet were arranged in tandem and were used mainly for simulated instrument flying; that is, making actual flights, but with the student's having no reference to anything outside the cockpit. On takeoffs, Burg would climb very steeply and then make some acrobatic maneuver, a maneuver that a fellow pilot in the back seat for simulated instrument letdowns did not always appreciate. Nor was that appreciation heightened in view of later events. As the years have passed and a number of serious accidents have occurred, commanding officers have been directed to crack down on aviators, often excellent pilots, who have engaged in extreme, unnecessary maneuvers. One recent incident involving a high-ranking Air Force officer flying a B-52 out of Fairchild Air Force Base in Spokane, Washington is typical. Known as an excellent pilot but inclined to rather extreme and unnecessary maneuvers to showcase his skills, his excesses were common news. They were resented by other pilots who thought them too risky; yet no one clamped down on him. In his case, he put his B-52 Stratofortress into a very steep turn low to the ground. As mentioned elsewhere, there is no vertical lift in a ninety degree banked turn and only slightly more in other steep turns, requiring that the pilot level his wings and perhaps add engine power to obtain vertical lift. But in that steep turn, the aircraft is losing altitude, and it requires some seconds to level the wings, sometimes there being not enough time, airspeed, and altitude to counteract the descent. As easily imagined in this case, a senior pilot and an even more senior pilot were killed, and a hugely expensive aircraft was destroyed — and the career of the commanding officer, who didn't do enough to control his rogue pilot, was likewise destroyed through disciplinary action.

It would be tempting to call Stephen Burg a "hot shot", or, in today's vernacular, a "hot dog." However tempting that may be, it would be an emotional rather than a rational appellation, reducing to the simple a matter of some complexity. Burg was very good, very daring, and seemingly very stable, or at least his confidence would make him seem so; and no one doubted that he had guts. Nor did anyone doubt that he was intelligent. His manner and speech were self-assured and convincing, although some had the feeling that, for whatever reason, he went about trying to impress people. He really did not have to. Though hardly what you would call arrogant, he somehow conveyed the idea that he was superior, reflecting this impression through seeming condescension toward officers of his rank and lack of respect for those of higher rank. He seemed to have the aggressive, fearless qualities necessary for good fighter pilots. Being overly cautious, overly safety conscious can adversely restrict those qualities. But a balance does have to be struck so that the pilot not do foolish things. This matter of aggressiveness, a good quality in combat, versus foolishness is a source of continuing discussion among aviators, a source hardly necessary for elaboration here.

It may be that Burg's confidence precluded any thought of problems or accidents. Young military pilots, like healthy people, think that bad things happen to someone else. However that may be, his youth and aggressiveness no doubt contributed to his feeling of overconfidence or invincibility on a weather flight one morning when, after completing his observations of fairly good weather conditions in the area, he decided to engage in an aerobatic maneuver or two before letting down and landing. Burg was flying a McDonald F2H Banshee, a two-engined jet of good reputation, and an aircraft in which Burg had lots of experience. There was a difference, however in this Banshee, because the model was an F2H-2P, the P designating a photographic version used for terrain mapping and other suitable military purposes. The F2H-2P, with its cameras in the nose of the airplane, affected the balance and drag characteristics to a slight degree — apparently being contributing factors in creating a problem in Burg's aerobatic maneuver.

Being at high altitude, Burg pushed the nose down to pick up the speed necessary to do a "loop", a complete circle, at the top of which the plane is inverted. Continuing, the plane then starts its downward path, ending up at the bottom in level flight to complete the loop. On this day and at some point approaching the top of the loop, the aircraft stalled and started to spin; however, it was not the normal spin, one in which the pilot is seated in his seat, but an inverted spin, one in which the pilot is upside down. In the maneuver Burg was attempting, any resulting spin would most likely be a normal spin which Naval Aviators are equipped to handle. Inverted spin recovery is not taught, inverted spins being more the provenance of aerobatic pilots flying smaller, lighter, more maneuverable, less sophisticated aircraft — even if recovery from an inverted spin is not necessarily difficult.

At high altitude, on his back in an inverted aircraft whose balance was different, and falling rapidly, Burg, in the unexpected confusion of the moment, had little choice but to eject. His ejection seat worked as advertised, his chute opened properly, he landed in a field and was picked up and returned to base. He had had none of the complicating factors associated with Tom Trundel's bailout, mainly because he bailed out at a much lower altitude and was not over water. The aircraft and its valuable photographic equipment were lost. Squadron pilots were usually not privy to all the details of the accident reports of other pilots in the squadron; and, for some reason — perhaps consideration of a fellow pilot's feeling — they did not ask questions which might further upset his equanimity. As to the loss of the aircraft, and without considering the conditions the pilot created, the model was being phased out of use, giving way rapidly to a much different, larger, improved version, the F2H-3.

The incident did not seem to affect Burg's confidence; but maybe to compensate, he continued his ways, subsequently being involved in a potentially serious, but, as it turned out, a much less destructive incident involving an F7U Cutlass, raising again the thought of "hot dog." And in fairness, though the error was mainly his, there was an external factor, a hazard which has been almost totally removed today. At an air show and flight demonstration at an East Coast Naval Air Station, Burg was flying an F7U Cutlass in rather toned-down maneuvers. The attraction of the F7U Cutlass for aviation enthusiasts was in getting a look at a very new, rather unusual looking aircraft with its long nose wheel strut and its cocked-up appearance. Some felt that it resembled a praying mantis. After performing routine and unimpressive maneuvers, it appeared that Burg felt compelled to impress the many people at the air show by planning his landing so as to touch his Cutlass down precisely on the very edge or beginning of the runway, even though there was no need for that. While such a maneuver seems risk-free or relatively risk-free, seasoned aviators know that, in flying, nothing is completely risk-free.

It was a combination of two things that would cause Steve Burg a problem. First, instead of landing at the very beginning of the runway, Burg miscalculated and touched down a bit short of the runway. In and of itself, landing a bit short of the runway need not necessarily cause a problem, because most all runways at major airfields have overrun areas that extend beyond both ends. Sometimes these areas are paved, and sometimes they are just grass areas. As luck would have it when things had not been going his way, particularly in view of his recent F2H-2P incident, not only was the area short of the runway a grass area, but it was also about six or eight inches lower than the hard-surfaced runway level. The difference created a concrete rise like a step at the beginning of the runway. So when the small, twin nose wheels at the bottom of the long nose wheel strut contacted the edge of runway at his relatively high approach speed, the support which keeps the nosewheel strut rigid broke; and the long strut and wheels folded back under the fuselage. As the aircraft rolled up the runway, the metal nose wheel strut left a bright trail of sparks but no significant damage. Removing Burg from a flight status was not considered, but few, if any, knew how much of a dressing down, tongue-lashing, or reprimand Burg received for the two accidents he caused by bad judgment. Ever a factor in aircraft accidents, however, is the accident investigation and assessment as to cause and blame; and no one likes to have himself receive the judgment of "pilot error." But it is safe to say that Steve Burg, intelligent and capable as he was, became a bit wiser and less self-assured, not to mention more mature and safer. It is also reasonable to remark that, whatever Burg's need to impress, his rogue behavior placed his usefulness to any serious squadron commander's plans into question — not to mention his own future. Accidents are one thing, but unnecessary accidents caused by reckless or grandstanding maneuvers are quite another. LTJG Steve Burg was a capable pilot; and if he could learn and mature from his mistakes, he would certainly become a better pilot. Whether he became a better pilot or whether he left the Navy is not known. One can only wish for the first and hope that he remained in the Navy. With maturity, he would have much to offer.

* * * *

In contrast to Burg, an equally good pilot flying an older, propeller-driven aircraft, the SNJ Texan, had no such personality hang-ups. Completely relaxed, down to earth, unpretentious, and completely likeable, Ensign George Mohr was the kind of guy who would have made a good combat pilot. But that was not to be. He finished flight training near the end of World War II; and much to his dislike, he was sent to Instrument Flight Instructor's school, after which he spent the next two years first at Pensacola and later at Corpus Christi, Texas. Having survived "the purge" of the summer of 1944, when the Navy realized that there were too many student pilots in the pipeline and decided to cut back drastically, he had hoped for fighter pilot training. With a zest for life and fun and being a good pilot to boot, he took his instructor duties seriously and worked hard at his job, always being ready for another flight. He was ever upbeat and cheerful, always being able to see the funny side of things. Having been born and grown up in Miami, he had developed a strong interest in nature and its creatures. Expert in diving, he had acquired an extensive knowledge of sea life, often telling stories about his successful searches for the many langoustes, which he loved to prepare for his family and friends. Equally knowledgeable about small animals, he once climbed a tree to capture a Racoon, bringing the creature down but acquiring many, many scratches on his arms and chest. But his real interest was in reptiles, about which his skill in handling water moccasins and rattlesnakes, made you feel there was nothing hazardous in the process. Aside from cooking a water moccasin during a survival training course, he had no desire to kill snakes. No matter where he was, he seemed to find others interested in snakes, often capturing rattlesnakes for those who then provided them for study and research. His being stationed at the Training Command's satellite fields, Whiting Field in the Florida Panhandle and then at Kingsville, Texas, not too far from the Mexican border, he was able to continue his interest in nature in his free time. It was all that he could ask for: the pure joy of flying combined with the pursuit of his hobby in the best possible geographical areas.

Though he was a professional in his approach to his duties, he longed for more excitement in flying. To compensate somewhat for the sameness of each flight, even though there was a syllabus outlining the instructional steps, he wanted to deviate from what is essentially "straight and level flight" by doing a little acrobatic work at the end of the last instructional flight of the day. Never using much time and never shortchanging his students, he wanted to take a few minutes for the acrobatics he enjoyed including spins, slow rolls, snap rolls, loops, chandelles, Immelman turns, and what not, to demonstrate to his students just exactly how those maneuvers were to be executed in the Texan, a two-seated, intermediate trainer, secondarily, but well suited for simulated instrument training flights. How many times Mohr engaged in his love of acrobatics was not known; and he was not about to broadcast the information, although it seemed that it was a daily occurrence.

Not a careless pilot, but nevertheless a confident pilot, he did not expect to have a problem when he once more demonstrated a spin, a spin he had handled time and time again. Expounding on the fine points of the spin to his impressed aviation cadet student, he encountered something he had never experienced before. Having raised the nose of the aircraft to the stalling attitude, he kicked the left rudder and the aircraft went nose down into the spin as it had on numerous occasions. But this time, the usual application of opposite rudder and forward pressure on the control stick was completely ineffective. He appeared to be in a tight spin, one against which nothing seemed to work. Usually, he would enter the spin procedure at six or seven thousand feet, although it was not a great height, but enough for the expected routine spin. Certain that he had sufficient altitude to pull out of the spin at a safe level, at first he was not concerned at the lack of responsiveness. But he was losing valuable seconds before deciding what to do after several corrective attempts to bring the aircraft out of the spin failed. With precious little time left to determine whether he would be successful, he told his student to bail out immediately. Then George Mohr unbuckled his shoulder and seat belts and prepared to follow. As George got himself hunched up on the seat to go over the side, he thought, "If I cream this airplane, they'll really put the screws to me." With that, he sat down; and without strapping himself in, he made one more attempt to come out of the spin. To his surprise, this time the plane responded beautifully to his use of opposite rudder and forward pressure on the stick. He was below five hundred feet when he pulled out completely and would not have made it had he attempted to rebuckle his safety belts. Instinctively, Ensign Mohr immediately climbed to a higher, safer altitude to look for his student, observing his student still much above him but descending gracefully in his parachute. Mohr thought how good it was that the aviation cadet had not delayed but had bailed out instantly on command. After watching his student descend safely to the ground and getting a good fix on his position, he called the air station to report his student pilot's location. George reluctantly headed back for the air station, deliberately postponing and fearing the moment on his return when he would receive some sort of official reprimand or at least a severe "ass-chewing." But since no one was hurt or injured and George had done nothing illegal, nor had he destroyed an aircraft, the matter was not considered worthy of censure and was dropped.

The bailout provided interesting fodder for squadron conversation, most of which centered around why the aircraft did not respond to corrective action when two pilots were aboard and why it did respond with only one pilot aboard. Some thought that at first the aircraft was in a relatively flat spin and harder to control, and that the student's bailing out affected the center of gravity, making it then easier to control. Others felt just the opposite: that the aircraft was at first in a very tight, rapid spin and that, after the student bailed out, the spinning became less tight and more manageable. Without stating what seems obvious, most seemed to agree that the student's bailing out was definitely a factor. The discussion would go on and on. Whoever was right, most pilots thought that George was at least half right, salvaging the situation but at very great risk to himself. Considering the time of this incident, it was probably a sign of an earlier, more relaxed but still demanding time, that the command did not make a big thing of the incident. Some in the military, and maybe many not in the military, might think that Ensign Mohr's actions warranted some sort of official reprimand. In today's litigious society with its "cover your ass" mentality, it is not hard to believe that such a reprimand might be forthcoming. But to this author and pilot, it is refreshing to think that such a time existed. Besides, an eager and daring pilot, like so many who daily put their lives on the line, Mohr had exceptional ability to be a dedicated mission-oriented pilot and exceptional combat participant.

While Ensign George Mohr was a good pilot, never having had an accident in four years of flying for the Navy, he experienced hard luck in flying civilian aircraft after he left the Navy. Waiting for an assignment to become an airline pilot, he did some instructing in small aircraft. On one occasion while instructing in Miami and sitting in the back seat of a two-seat aircraft, his airplane and another two-seater collided in midair. The planes became locked together and came down from several thousand feet twisting as a Maple leaf rotates from the branch to the ground. The impact was severe, trapping and killing three of the four persons involved. Only George Mohr survived. And unfortunately several years later, this fine young person and aviator was killed in an airline accident while he was flying copilot. It might be easy for some to say that George was jinxed — or even destined to die in an aircraft accident. But no one who knew him could possibly think that. He loved flying and he loved life and would do nothing to even slightly damage an aircraft, or to shorten his flying career — or to end his life!

Part Five: Formation Flight

Formation Flight — A Means to an End!

The importance of formation flight in military aviation is often overlooked and little understood. Other than military pilots, few probably give it much thought; but so much is accomplished in formation flight of one kind or another that that is really "where the action is." Aside from the beauty and precision of a large flight, the formation represents a means to an end. Carrier fighter and attack aircraft are almost always in a formation, whether it is the smallest unit, the "section," which consists of just two aircraft; whether it is a larger unit like the "division," which usually consists of four aircraft; whether it is the larger unit, the "squadron, which is made up of all the divisions in the squadron;" or whether it is the still larger unit, the "air group," which consists of all the squadrons making up the air group. Fighter aircraft would naturally like to operate in sections for combat because of greater flexibility available than in larger groups. The well-known "Thatch Weave," named after its Naval Aviator inventor, which provides greater advantages in combat, utilizes aircraft in two-plane sections. Depending on the mission, several divisions or even the entire squadron of four, five, or even six divisions could be involved. Over the years, the number of aircraft in a squadron has changed. Just after World War II, a typical carrier squadron would have as many as twenty-four aircraft; and the next larger group, the carrier air group, would have as many as one hundred twenty aircraft assigned, a number that could be accommodated by a single carrier of the Midway class. The typical air group then would have three fighter squadrons and two attack squadrons. More recently, as carrier aircraft have become bigger, heavier, more sophisticated with high tech electronics equipment, and more specialized duties, the number of aircraft in a squadron has decreased. And even as aircraft carriers have become larger, they accommodate only about eighty to eighty-five of these aircraft today.

At one time, pilots in training, in addition to normal formation flying, would practice four-plane division takeoffs and even, less commonly, four-plane division landings in formation. Whether the formation landing practice had much to commend itself, particularly for young pilots with relatively few hours of flight time, is debatable. A pilot in training making a formation landing when he was on the left inside of the leader during the left approach turn to the runway was always somewhat apprehensive, mainly wondering whether the inexperienced lead pilot would take into consideration the "insider's" slightly slower, more difficult position when he was not too much above the stalling speed. Runways are hardly wide enough for today's larger, high performance aircraft to engage in more than two aircraft formation takeoffs and landings. Also, with the higher risks involved and the questionable advantages to be had, it makes little sense. In earlier times and in smaller, slower, and less demanding aircraft, formation landings were not uncommon — and particularly at training bases where the fields often had a "mat" landing area, a huge paved-over airfield. Regardless of whether it was runways or mats the pilots were approaching, the pilot on the inside of the approach turn to the runway had to fly slower than the division leader to keep from overrunning the leader, making him feel uneasy every time he had to throttle back a bit and giving him concern that he was close to the stalling speed of his aircraft, the speed below which he could not stay in the air. The pilot on the inside of the approach turn to landing was in a difficult situation: flying in a step-down position, he could not add power to get out of the formation for fear of running into the leader or losing sight of him; and if he throttled back any further, he would have to increase his angle of attack without adding power to increase his lift, a move that would have equally unfortunate results.

In the course of a year, the public will hear or read several times about military pilots bailing out or being killed in accidents while flying a "routine training mission." Indeed, they are routine missions; but often the accidents are the result of midair collisions while flying in formation flight. What lapses or freakish circumstances cause these collisions may not always be known, but every military pilot is aware of the hazards and potential for disaster which can occur during formation flight. Yet, despite good, safe procedures which have been instituted, problems can and do arise. The so-called "step-down" procedure is the standard in military formation flying, largely because the pilot has an unobstructed view forward and up, but may have an obstructed view down because of the low-wing design of the aircraft. With the swept wings of jet aircraft, the wings were largely behind the cockpit, so the pilots had a better view down. But nevertheless, step-down position flying is the usual mode in formation flying. The step-down position should minimize the danger when one pilot overruns the pilot ahead of him, enabling him to slide under the overtaken aircraft with no harm being done usually, other than the creation of momentary confusion and his losing sight of the aircraft ahead. But formation flying requires constant attention and concentration — there is no doubt about it!

While the number of formation flying accidents remains small, they do happen occasionally — and usually with disastrous results for one or both pilots. An unfortunate training command accident emphasizes the risks and consequences. Two student pilots were returning to a Pensacola Naval Air Station field when the wingman slightly overran the leader, damaging his tail. The aircraft only touched momentarily; and afterwards, both were still flying in what seemed a satisfactory manner. Sensing no real or imminent danger, the pilot decided to make his landing approach. He lowered his wheels and then wing flaps for landing, requiring the adjustment of his horizontal stabilizer (elevators or horizontal tail surface), slightly altering the aerodynamic forces. Additional forces were placed on the tail surfaces with the result that the entire tail broke away from the airframe, and the aircraft dropped nose-first into the ground with the pilot's having insufficient time to bail out. In retrospect, a more thorough testing of the aircraft after the collision may have helped; and even if it was deemed safe for a landing approach, a flaps up landing would likely have placed less stress on the tail. Perhaps a more seasoned pilot would have been more cautious, but who can really tell?

Much less serious situations in formation flying occur, which are more comical than dangerous. In their step-down formation, it was not unusual for AD Skyraider pilots in a division to scatter in all directions as the lead pilot, or another pilot in the division, would run dry a droppable, one hundred fifty gallon external gasoline tank after about an hour's flight. There would be a puff of black smoke, a decided slow down, and the beginning of a descent. There were no gauges for the pilot to know exactly how much fuel was left, but the pilot would know immediately what his problem was and switch to the main or another tank; the division would then regroup and continue its mission. The absence of a fuel gauge caused this situation to occur occasionally, and no one gave it too much thought, despite the potential for midair collisions. Night operations have produced their own share of incidents, as described herein later. One disadvantage of flying in close formation is that, for other than the leader, it is very difficult — often impossible — for the wing men to dial their radio and other equipment for navigational purposes, thereby relying almost completely on the leader to determine the formation's position. That is not to say that the leader can be carefree, because he must think ahead about changes to be made in throttle settings and in making turns. It takes but little imagination to think what awful things can happen otherwise.

The Air Group Goes Aboard Ship: Patience Required

The Operational Order (op order) had been drafted and completed, its significance being that the entire air group would fly aboard the *USS Midway* (CV-41) for a four-week period to conduct various training and qualification exercises. This meant that all five squadrons of the air group would leave the Naval Air Station Norfolk, Virginia, rendezvous nearby and fly out to the carrier in one large formation — and what a sight it was to see one hundred and twenty or so propeller-driven aircraft in formation. This formation experience would provide valuable training in many ways: in addition to the practice of formation skills, it was a lesson in team work and discipline, exercised for the good and pride of the whole unit; it provided opportunities to both lead and to follow someone else's lead; and the stimulation received helped to foster a desire in each pilot to do his best.

If you had not experienced it, you would be hard-pressed to believe that such large formations of naval aircraft ever existed. In today's Navy of supercarriers and large, high performance jet aircraft, large formations are just not very practical — and certainly they are not very efficient, certainly less efficient than with propeller or "prop" types. Today's shipboard complement of about eighty aircraft contains a mixture of fighters, attack aircraft, and a small assembly of aircraft for specialized missions like airborne early warning radar aircraft, etcetera.

But, before the switch-over from propeller aircraft to turbo jets, A Midway class aircraft carrier *USS Midway, Roosevelt, Coral Sea* (CVB's 41, 42, and 43) would usually have five squadrons of from twenty to twenty-four aircraft each, with three squadrons being fighters and two being attack.

Depending upon the air group commander's assessment of training needs, there would be times when the entire air group of one hundred twenty or more aircraft would rendezvous into one giant formation (See Illustrations). It is one thing to mention the figure of one hundred twenty, a figure which in and of itself may not sound impressive; but it is quite another thing to be a part of such a group or to view the formation from the ground or from an aircraft carrier at sea. Only a very small number of people have had an opportunity to see such a formation from the ground; and it is a much smaller group of men who have had the opportunity to be a part of that formation.

With this operation, as with all similar operations, decisions had been made about whether to load the aircraft on board with slings and cranes or whether to have the air group fly out to the carrier after it has left port. Support personnel, spare pilots, and any "hangar queens" with potential for shipboard repair had to go aboard before the carrier left port. Considering the demands of maintaining and inspecting aircraft, it is amazing how often all, or all but one or two, of an air group's aircraft are in an "up" status and ready to fly out to the carrier and land aboard in such situations.

Consider the decision to have the entire air group fly aboard. The three fighter squadrons take off first, followed by the two attack squadrons, all of which have to rendezvous according to an established procedure and manner. Although the takeoff interval between planes is very short, as short as fifteen or twenty seconds — just enough to partially clear the slipstream turbulence of the aircraft ahead — only three or four planes get airborne per minute. At the very best, and provided that that rate is maintained without interruption, at least thirty to forty minutes are required to launch the entire air group. While some aircraft are waiting for takeoff, those that have taken off are climbing to a prearranged altitude to rendezvous with the squadron flight leaders, usually the commanding officers, who are flying in a large circular pattern. Although the planes will fly in a close formation, about fifteen feet apart, each aircraft will be flying in a slightly stepdown position. This serves a two fold purpose: since most tactical aircraft are of low-wing design, the pilot has a better view looking slightly up than down; and also, there is less danger of a midair collision if, for some reason, one aircraft should overrun the aircraft ahead and lose sight of it under his wing. Each squadron will also be in step-down formation relative to the squadron ahead. Viewed from a horizontal position, or side view, the squadrons would resemble steps leading up to the first pilot in the first squadron. The air group commander as the senior officer and "Boss" of the entire air group is free to fly as and where he pleases to oversee the entire operation.

As exciting and stimulating as the thought of returning to sea and carrier operations are, the pilots in the attack squadrons know there will be a certain amount of tedium as they wait first for the fighter squadrons to take off before they can and later when they must patiently orbit the ship while the fighter pilots land first. No one was more aware of what tedium lay ahead than LTJG Wilson Baren, who had done this many times before. In consideration of the situation pilots faced, they would often rotate the fighter squadrons so that the same pilots would not always be the first to land aboard. The attack squadrons would also be rotated, but they would always be relegated to the same fourth or fifth squadron location.

After the air group had completed its rendezvous, it headed out to sea to meet the carrier, which, by custom, was not much more than an hour's flight away. Once the air group reached the vicinity of the carrier and had received a "Charlie" to begin landing, it started the process by flying along the starboard side of the ship; and once past the bow of the ship and already having received a "Charlie" sign or green light to land aboard, the lead squadron broke off to the left and start descending, while the rest of the air group would start a level left turn shortly later. The lead squadron, descending as it turned made a wide and complete turn to fly along the starboard side again. By this point, the divisions in this first squadron had moved over to the right of the lead division in a right echelon. This time, as they reach the bow of the carrier, the pilots in the first and second divisions peeled off individually to the left, creating an appropriate interval between aircraft for efficient landing intervals. As each individual pilot broke off, he descended farther, went over his landing check list, including the lowering of his tail hook, and then leveled off at two hundred feet above the ocean. He flew downwind and parallel to the carrier until he was about opposite the stern of the ship. At that point, he started a left approach turn to make an arrested landing. The remaining aircraft in the first squadron maintained level flight until their turn was completed and then followed the routine of the first two divisions. Meanwhile, the other four squadrons above waited their turns to break off, descend, and then break off individually for landing. Several times LTJG Baren had been in the sixth division of the fifth squadron to go aboard, so he well knew how long and patiently he would have to wait. But he was no different from the other pilots in similar positions who knew about time and patience also. It was the intense concentration of formation flying and the constant throttle adjustments to keep in position that constituted the hard work that lay ahead and which tended to dull the romance of the whole operation.

The entire procedure was designed to be as safe as possible. A twenty-five second landing interval between aircraft was considered excellent — and so it was. While this desirable landing interval was often maintained for awhile, needless to say it could not be sustained while so many aircraft were involved. On touch down on the carrier deck, the aircraft's tail hook would engage an arresting wire; and the aircraft would be brought to a stop. After landing, the pilot would let his plane be pushed backward by the force of the wind across the deck until the arresting wire dropped free of the tail hook. The pilot would then raise his tail hook, and taxi forward rapidly at the direction of the flight deck crew until he was past the barricades which protected planes parked forward. The landing portion of the carrier deck was then clear. All of this was taking place while the next aircraft was rapidly approaching the stern of the ship on its final approach. The disengagement from the arresting wire and the taxiing forward of the barricades could take fifteen seconds of the desired twenty-five second landing interval, so it is easy to see how close the next aircraft must be when the plane ahead clears the barricades. It is likewise easy to see how a minor problem in disengaging the arresting cable or in taxiing forward can cause the next aircraft to take a wave-off, doubling the landing interval — and somewhat backing up all the planes still to land. Other things like accidents cause delays as well, the result being that it requires longer than the sought-for ideal for the entire air group to land aboard.

With one hundred twenty aircraft to land aboard the carrier, a full hour would be required to get everyone aboard — even averaging a slightly higher, but still quite good, thirty second interval. No doubt on some occasions, an hour was all that was required. Most often, however, an hour and a half or more was required — particularly if there were an accident which required that the aircraft be hoisted away by a moveable crane or "Cherry Picker." Pity the last squadron, and particularly the last division of the last squadron. The pilots would have aching necks from having their eyes fixed on the division leader; and the individual pilot's left arm was fatigued from the constant throttle adjustments required to maintain his precise position in the formation. Theirs was a special pleasure when, after two and a half hours of flying, at long last, they were on the deck. Wilson Baren knew that other attack squadron pilots were as glad as he was to finally be aboard.

With the transition to turbo jet aircraft, such large formations became a thing of the past. Many adjustments were required in procedures, including a much longer landing interval.

Night operations to locate and land on aircraft carriers in the days of limited navigational equipment, that is without the precise navigational TACAN or automatic direction finding (ADF) equipment, would sometimes prove interesting. One example can be illustrative. On one dark, overcast night after eight pilots had made their night qualification carrier landings, they were advised to take off, rendezvous, practice night formation flight, and, in general, just get off the deck and out of the way to make things easier and less time-consuming for the other pilots to make their qualification landings. After about an hour and a half, the flight leader, Naval Academy graduate LCDR Wes Sabatino, was contacted by the carrier to tell him that he could bring his eight aircraft back and land aboard. He was told that he had a "Charlie" meaning he and his group were cleared to start landing. He acknowledged the "Charlie" and started back to the carrier — or where he thought the carrier was. Using his night navigational equipment, a system called ZB, he was, generally headed in the right direction; but he was farther from the ship than he thought. Receiving another call from the carrier that he had a "Charlie," being certain that he was close, and anxious to exhibit his leadership, he led his group down alongside some ship lights, and when just past the lights, gave the signal to break off individually to commence landings. In all fairness, and on such a moonless, starless night, no one else in the flight could be certain whether it was or was not the carrier. Other pilots in the group might have had their doubts; and, according to accepted procedure they were obligated to say something if they thought that the leader was about to make a serious mistake or endanger the formation. But they were pretty well occupied with the concentration and effort required to fly in formation at night to challenge the leader's judgment. It was only after the pilots broke off individually that their doubts grew. As the first to break off, Wes, the flight leader, was the first to make his approach turn to land. He was more than halfway through his approach turn when he realized that the ship's deck-edge lights would not come on and that it was not the carrier but some fishing or commercial vessel. He called for the others to join up, using some terms to mask his call from the ears of the people on the carrier. Since the carrier was nearby, it was not too difficult then to find the real carrier. You can be sure that there were some caustic and humorous comments floating around for some time after this night. While the flight leader was mostly responsible for this mix-up which was not serious and which had a happy resolution, the pilots knew that they had an equal responsibility to speak up if they thought the leader could be making a mistake in navigation.

During World War II, and in stressful conditions of low fuel states and black nights, there were many instances of pilots trying to land on ships they thought were carriers, and some landed by mistake on the wrong carriers, thinking they were their own. If the reader thinks these types of incidents just should not happen, consider that at night pilots have landed at the wrong airports. Airline pilots have made this mistake, one even landing at the wrong airport in daylight. The lights on carrier decks at that time were quite dim, and in the case described above it was possible that the ship had not yet turned on its deck lights. At best these lights, called dustpan lights at the time because they resembled nothing so much as inverted dustpans, directed their dim lights downward just over their own locations at the port and starboard edges of the deck. And strange as it may seem to some, carrier aircraft had no landing lights. The pilots making approaches had to line up between the deck-edge lights and estimate their height above the deck. Even at the Naval Air Stations, the same procedure obtained.

History buffs who have had an interest in aircraft carrier warfare in the Pacific Theater during World War II will no doubt know of the famous story of Naval Aviators returning to their aircraft carriers low on fuel and after darkness had set in. It was on June 19, 1944, in the Marianas Campaign, when United States Forces were getting ready to invade the islands of Saipan, Guam, and Tinian. Commander Edward P. Stafford, USN (Retired) describes the setting in his book *The Big E* (the aircraft carrier *USS Enterprise* CV-6): "But Japan could not afford to lose the Marianas. They were too close to the homeland, within heavy-bomber range, and U.S. naval and air forces based there could cut off everything to the south. The Combined (Japanese) Fleet, which had not been committed since Midway, was ordered 'to attack the enemy (U.S. Forces) in the Marianas area and annihilate the invasion force.' But (Japanese) land-based air was ordered to destroy 'at least one third of the enemy task force carrier units... prior to the decisive battle.'" At the same time U.S. Naval Forces under the command of Admiral Raymond L. Spruance wanted to engage the Japanese in advance as a means of protection for his invasion force, his primary mission. Without going into too much detail, it was the afternoon of June 19 when, over Guam Navy Hellcats (F6F's) shot down so many Japanese aircraft that the air battle was called "the great Marianas Turkey Shoot."

Of equal importance was Admiral Marc Mitscher's decision. At 4:10 PM on the same day, June 19, 1944, on hearing about the location of this large Japanese fleet at about the maximum range of his aircraft, he decided to launch his aircraft from the carriers under his command. "In the fifteen minutes between 4:21 and 4:36, Mitscher's carriers turned into the wind, launched 54 torpedo planes, 77 dive bombers, and 85 fighters and turned back to the west again. It was not a routine task but a desperate, last-chance effort to destroy an enemy who for two days had remained maddeningly out of reach, and was even now at extreme range and capable of escaping entirely unless this opportunity were seized."

The essence of Mitscher's decision was that the pilots would be returning after dark, since they did not depart on their mission until 4:45, and "it was a minimum of two hours out, two hours back and a fight between — and sunset was at seven o'clock. Two hundred and sixteen planes would have to land in full darkness with nearly empty fuel tanks, many with battle damage. But it was his only chance to strike the enemy and Marc Mitscher had no choice." It was Admiral Mitscher's tough decision, but it was the unquestioning pilots who guaranteed its success by attacking the Japanese Fleet and damaging the aircraft carriers *Ryujo*, *Junyo*, and *Hiyo*, knowing full-well that some would not return and that those who were still able to return would not only have to find their own carriers again but also contend with low fuel states and possible unknown damage.

Finding their way back to the carriers after dark was no sure thing, not to mention their landing aboard ship when very few had much experience with night landings on aircraft carriers. Here again it is best to use CDR Stafford's words. "After the long flight, after the AA and the (Japanese) Zekes, now in the darkness, nearly out of fuel above the empty mid-Pacific, many of the pilots became desperate, and the air was full of calls to anyone for help. 'Can anyone tell me where I am?' 'Give me a vector. Can anyone give me a vector?' Others were calmer."

"Hey, Joe how much gas you got left?"

"About a Gill." (A quarter of a pint)

"I've got about two gills. How about you, Tom?"

"It reads five gallons less than empty."

"Let's put 'em down while we still have power."

"And the planes began to go down. The pilots realized that their chances of ditching successfully at night were far better if they still had power with which to control their planes, and, when the fuel gauges read zero and no ships were yet in sight, they began to land. One group of SB2C Helldivers ditched in formation and lashed their rafts together. But most continued eastward." At about forty-five miles, some of the aircraft could receive the carrier's radar beacons. "The biggest disaster in the history of aviation was in the making. The pilots needed lights. But there was a war on in the Pacific. Enemy snoopers were in the air. Enemy subs had to be around. There were between 300 and 400 men total, in the circling 200 aircraft overhead and inbound. One torpedo spread could sink a carrier with 50 planes and 2,500 men aboard. Logic demanded that the ships stay darkened and the airmen land as best they could. Destroyers could pick up most of them in the morning."

In this difficult situation, Admiral Marc Mitscher thought hard about the risks, realizing that he did not have the luxury of time to decide. He turned to Captain Arleigh Burke and uttered the now-famous words, "TURN ON THE LIGHTS!." "Task Force 58 sprang out of the darkness as though a ceremonial master switch had illuminated a Worlds Fair. Thirty-six inch and twenty-four inch searchlights poked bright, vertical fingers into the sky. Red truck lights glowed in pairs on the mastheads. The carriers turned on their deck-edge lights, and destroyers fired star shells which broke and drifted down, lighting up the sea for a mile in all directions. The urgent, empty planes began to land, but the lights could not make it day, nor put fuel back into empty tanks, nor even show which carrier was which... Confusion still ruled the night sky over Task Force 58. Planes cut each other out, crowding in to land on any empty deck. There were looming, rushing near collisions that left both pilots angry and shaken. Planes made approaches to battleships and destroyers. In a hurry, there were bad landings which fouled the desperately needed decks so that planes, which could otherwise have landed, ran out of fuel and ditched."

Space here precludes a further description of the events on the night of June 19, 1944, other than to note that one can only imagine the consequences had Admiral Mitscher not turned on the lights. The decision to launch an attack on the combined Japanese fleet so late in the day brought with it huge risks, all of which had to be considered — and none more so than bringing so many aircraft so far home, after their attacks, to the darkened aircraft carriers with so little fuel. But much is owed to the pilots, who did so much destruction to the Japanese carriers and then had to find their way back with very limited navigation equipment, nearly exhausted fuel tanks, some radio silence, and darkened ships. CDR Stafford, in his book *The Big E* does not go into detail about exactly how dark the night was and how restricted the visibility was. But it is reasonably certain that it was not a bright moonlit night when the aircraft carriers and their escorting ships were lit up by a full moon and bright stars. With a full moon and good visibility, the moon would be reflected off the ocean mirrorlike, and the ships could have been seen for miles. On such bright full-moon nights with water reflections, junior officers like to say jokingly that on such good nights even the commander would fly, giving rise to a persistent appellation of "a commander's moon." In reality, no one could say there was any truth in such a reference; but to this day, such beautiful full moons on the water at night are referred to as "commanders' moons."

The Carrier Returns from Sea: Launch the Airgroup

By this time they were ready to get off the aircraft carrier. It wasn't that the pilots found air operations from the ship tiring or unpleasant. On the contrary, these operations, if sometimes with unanticipated consequences, were stimulating and satisfying because all of their efforts contributed toward the attainment of further qualifications and improved readiness. It was just that, now that things were winding down with little left to be accomplished, everyone was looking for a change, a change that would reunite families and provide a period, however brief, of more normal living.

Yet most of the Navy people aboard ship were not unmindful of the many enriching experiences of beautiful sights and interesting places they had visited, places that they had only heard of or had possibly read about in their still-young lives. Every period of duty aboard ship provided fond memories. On "cruises" like this one they had seen much. In the Caribbean, they had, quite naturally seen the Naval Air Station Guantanamo Bay, Cuba, a base on the south shore and toward the eastern end of the island, which had been leased to the United States for one hundred years following the Spanish-American War. They could not forget those motor launch liberty trips from the carrier anchored in Guantanamo Bay to the Fleet Landing points in the inlets at night. Those trips seemed magical to practically everyone, not just the romantic, young crewmen and pilots. Once into the inlets and lagoons, the water was so smooth and peaceful, the only sounds being the boat engine's now slow propulsion and its rippling water-effect; at the same time, the mirrorlike surface of the water reflected the dim lights in the vicinity, and the balmy soft breeze that came with the evening hours combined to compose an almost magical setting which no one ever tired of.

The air even seemed perfumed as if a fragrance of the blossoms intensified at night. The young, barely out-of-their-teens, pilots, headed for their goal, the officers club, and the newly discovered Añejo Rum Punch and Cerveza Hatuey, drink names which would be forever remembered. But the Naval Base was not the only place they would see in the Caribbean. Most went ashore in Puerto Rico, Haiti, the Dominican Republic, and Panama, with still others, usually pilots, seeing a small part of Jamaica. And many of these same people would also cross the Atlantic with the aircraft carriers to awake one morning to the thrilling sight of the pillars of Hercules, with Gibraltar on the north or left and the high Atlas Mountains of Morocco on the right or south, which together formed the gateway to the Mediterranean Sea. Most of these young Navy people had only seen pictures of the Rock of Gibraltar in advertisements of Prudential Insurance Company; and soon with the carrier tied up to the pier, they could get off the ship and explore the Rock, hoping to see some of its famous monkeys. Later, with the carrier at anchor in Gulf Juan, they would have the opportunity of a day's liberty in Nice; and still later, sailing through the Strait of Massina and viewing the smoking Mount Etna, and the ship's tying up at the pier in the Bay of Naples, they could visit Pompeii or even go to Rome for an audience with the Pope. Some even had a day's liberty in Tripoli — yes Libya. Pilots, of course, had the better views, their missions having them fly close to or over Sicily, Crete, and Malta, and even making bombing and strafing runs on Filfa Rock just west of Malta.

Each period at sea — or "cruise" — brought many experiences, many memorable for the opportunity to see new things and, unfortunately some unpleasant things involving injury or death. They would put aside the unpleasant things because some occurrence of that nature accompanied all periods at sea; and death, however infrequent, was a fact of life. No one was unappreciative of the pleasant and uplifting things he had seen. But all periods at sea had to have an ending, and everyone seemed ready for a change. The air group's pilots knew they would be the first to leave the carrier, a privilege that came with their function; otherwise, the aircraft would have to be off-loaded at the pier at the Norfolk Naval Base, the off-loading being a very time-consuming process. The pilots might be back at Naval Air Station Norfolk a full day before the carrier tied up to the pier. LTJG Harry Holgate was typical of the young pilots. Having had several cruise periods, he remembered the way he felt when the air group was preparing to fly aboard. He was excited as usual, feeling it would be good to go back aboard ship, to become a part of the 3,500 man complement of trained and efficient specialists of every description whose only purpose was to maintain and utilize the aircraft in the air group. That feeling was the plus side. But moving about the ship through the narrow passageways, going through shin and knee-banging hatches and watertight doors, climbing stairs so steep they were called and resembled ladders, sleeping on canvass bunks in twelve man compartments, and using big washroom-toilet facilities, while not really what you would call primitive, did not really constitute comfort — even for young men. After a week or so of being back aboard ship, many felt that they had never really left the ship after the last cruise. Most of all, even though Holgate could endure anything involving the carrier, these minor but daily inconveniences heightened his anticipation about flying off the carrier now. He was no different from most of the pilots who, though anxious to fly off, were willing to go back again when required to carry out their duties and responsibilities.

LTJG Harry Holgate had been fortunate enough to be scheduled to be included as part of the departing air group. Most pilots would be scheduled, but there were always several who could not be included because there were a few more pilots than aircraft; and there were usually a few aircraft in the air group that could not be flown because of needed parts or maintenance. Thinking about the launch about to commence, Holgate thought how amazing it was that nearly every one of the one hundred twenty aircraft in the air group was ready to fly and that the launch of all aircraft would go off so smoothly and quickly. It was not chance but training and discipline that would insure such a coordinated and efficient departure. Harry Holgate, along with many other pilots, had been a part of many of these group launches and had only seen one mishap: This fighter aircraft, an F8F Bearcat, whose history of mechanical problems justly earned it the status of "hangar queen," had been brought up from the hangar deck on the deck edge elevator to take its position in the sequence of the last fighter squadron. Holgate, located near the front of the first attack squadron, could see clearly the fighter pilot taxi into position and take off. Still fixed in position because of the few remaining fighters to take off before his squadron could move, Holgate observed the Bearcat climb to about three hundred

The Air Group as It Goes to the Carrier or Returns

196   *Open the Hangar Doors*

feet at about a mile directly ahead of the carrier and then slowly start to lose altitude. The descent appeared in slow motion, and Holgate, hardly believing his eyes or thinking it would go down, could see it all the way to the water. No "Mayday" was reported, and Holgate wondered whether he should get on the radio to report it. Surely, he felt, someone on the bridge had seen it too, so he said nothing. Within a minute or so, Holgate was airborne headed toward where he expected to see a still partially-submerged aircraft or a pilot in the water with his inflated Mae West. To his complete surprise, since the water landing seemed reasonably gentle, he saw nothing of either, only an oil-free, smooth circle of water caused by the aircraft's sinking. As simply and quickly as that, a fine young man, a husband and father, was lost. The life of a man of good character, intelligence, and skills was lost "just like that" in such an unexpected way. He would be mourned by family, squadron mates, friends from other squadrons, and others. Most pilots in the other squadrons in the air group did not know him. No doubt, all the pilots in the air group had seen each other — particularly aboard the aircraft carrier — but each squadron had its function and responsibilities which precluded socialization on a large scale. It is very likely that, although every pilot in the air group was saddened, many had no idea which face went with a particular name. Holgate had heard of other Bearcat ditchings under worse conditions without any real problems; and he even knew of one in which, instead of going into the water in a wings-level attitude, had struck the water in an almost vertical, nose down attitude with the pilot's surviving. Under the circumstances, many questions naturally arise; but what they were remained the knowledge of his squadron, the air group commander, and the carrier's skipper. Other pilots were left to wonder what had gone wrong. Could it have been the cockpit canopy? It had been open for takeoff and would normally have been closed right after takeoff. Had he forgotten to open the canopy before ditching in the sea? He had so little altitude when his engine failed, and so many things had to be going through his head. He was headed into the wind so his landing would be at relatively low speed and ease the impact. Chances are he did everything right. But something went terribly wrong; and what looked as if it should have been relatively safe turned "wormy." You just never could predict. This upcoming launch, however, would have no such incident to mar it. Everything would go, as it usually did, without a hitch.

At the beginning of a "cruise" or departure from a Naval station aboard an aircraft carrier, whether for a short few weeks or whether for a six month's deployment overseas, there is always a special excitement about the entire air group's mission, because it symbolizes the start of something new and of all the events yet to occur. A related but quite different experience is found in the launching of aircraft from the carrier at the end of a cruise and their return to the home Naval air station. While the air group's flying out to and landing aboard the carrier is seen by just a few people, part of the carrier's crew, the return to home base is seen by thousands.

While Harry Holgate did not know it at the time but would learn later, many people like Joe O'Malley and his wife, Susan, civilian neighbors of his who were walking south on Hampton Boulevard in Norfolk, were hardly aware of a distant, faint noise. In fact, Joe said that he could not be sure just how long it took to realize that the first distant, hardly noticeable sound was something other than his imagination. But slowly, surely, the sound became real until there was a definite low-pitched roar of aircraft engines. It took awhile for Joe and thousands of others in Norfolk that day to find the direction from which the roar came. Now looking and checking several directions, they could see airplanes, airplanes, and airplanes coming into view, not being sure how many and being unable to count so many moving aircraft. Everyone, once hearing and then seeing so many aircraft in a beautiful formation, knew that it was an air group returning from sea. Many even knew that it was Naval Air Station Norfolk's Air Group Seventeen returning from sea. And what a sight it was! Close to one hundred twenty aircraft (See Diagram) were flying south parallel to Runway 18-36, the North-South runway. At the direction of the air group commander, the air group made one huge race track turn over the center of Norfolk. Only the blind and deaf out of the hundreds of thousands of people in Norfolk could fail to see this sight. Completing the first race track pattern and heading south again parallel to Runway 18-36, the first squadron broke off from the others to descend and come around in a smaller pattern so that, once opposite the upwing end of the runway the first eight aircraft could break off individually and turn downwind. Once opposite the downwind end of the runway, each pilot in turn would start his one hundred eighty degree approach turn to the runway and to landing. The air show continued for some time, as each squadron and its divisions repeated the carrier formation procedure.

Observers became aware that, as each moment passed, fewer and fewer aircraft remained in the air and that a magnificent show was winding down and coming to an end. While many observers on the ground saw the entire show, and while Henry Holgate felt an integral part of the formation, his concentration on flying and maintaining his wing position meant that he could only steal brief moments to observe the overall process.

What effect the return of Air Group Seventeen had on the people of Norfolk would be hard to know, the effect probably being somewhat different depending on age and other factors. The newspapers mentioned it, saying how much people were impressed by the precision and the demonstration and saying that many wished they could have been a part of it. Not specifically mentioned, but obvious, was that it was a reminder of what the Navy meant economically to this Navy city. Also, unknown was whether this exercise, so much in evidence to all of Norfolk, was planned with certain things in mind. If so, those things were the most distant things from the minds of the young pilots in the air group, who were so skilled and focused on their formation duties and so unskilled and uninterested in the political message, if any, intended.

Superchargers and Formation Flight: A Problem

This flight as part of the overall training exercise had been known about for some time; but just when it would be scheduled was not known. Considered a significant part of the overall training for high altitude work, no one, however, attached anything but routine concern for the flight and general safety precautions. If anyone thought about it in advance, it was probably just a few pilots, the rest feeling that the briefing before takeoff would be soon enough to think about it. After all, that was the way things were done, so there was no need to concern yourself too much in advance. On Monday morning that March day aboard the *USS Coral Sea* (CVB-43), which was in the vicinity of Guantanamo Bay, Cuba, the attack squadron assembled in the ready room to receive the briefing for the day's flight. In four other ready rooms aboard the *USS Coral Sea*, similar briefings would be taking place, because this mission was to include the entire air group. Ensigns John Ritter and Winston Gray, good friends and neighbors back in Norfolk, sat next to each other as they often did during the preflight briefings. The briefing officer reviewed the mission, going over in detail rendezvous procedures, tactical radio frequencies, etcetera, and, most of all, engine procedures to be used in this high performance exercise. Simply stated the flight was to have the entire air group assemble in formation and, on command, increase engine power to "maximum continuous power," a state meaning essentially what it says: the engine could be operated at this setting continuously with no damage to the engine. After that, the air group would climb higher; and, on command, each pilot would engage his supercharger to enhance engine power for higher altitude operation. Without superchargers, the aircraft would have a very difficult time even getting to high altitudes. It was the superchargers that enabled the B-17 Flying Fortresses to fly at high altitudes over Germany during World War II.

The briefing completed, the pilots went on deck to "man their aircraft" as the term was used. On signal, each pilot started his engine and awaited his turn for takeoff. The bright morning air was a little cooler than usual, as cool as it ever gets on sunny days in the Caribbean Sea area; and the visibility was unlimited. But as usual, a little breeze was blowing, but not as stiff as it would get later in the day when the heat was sure to build up. The carrier was now about fifty miles south of Santiago, Cuba, as most of Air Group Seventeen, having taken off from the aircraft carrier some minutes earlier, was completing its rendezvous just south of Santiago, Cuba, under the leadership of a short, tough, and competent air group commander. No one would think of disobeying or ignoring the air group commander. It just wasn't done, partly because of respect for authority and partly for his strength of character — but mainly it was because he was extremely capable and knew how to lead and motivate. The air group, highly disciplined and focused under this capable leader, consisted of three squadrons of F4U Corsairs and two squadrons of AD Skyraiders, both being good workhorse types of propeller-driven aircraft. Part of this training mission was to gain experience in operating at "maximum continuous power" or "military power," as it is frequently called. The other part required the shifting into "supercharger mode," or "high blower" as it is often referred to. Most likely, all of the pilots had shifted into high blower at some time or other; but since most flights were conducted at altitudes where supercharging the engine was not required, the pilots had very little experience in actually engaging the supercharger. Therefore, this exercise was considered a good refresher — talking about the procedures was fine, but there was really no substitute for actually going out and getting the actual practice. Essentially, the supercharger was used to compress air at higher altitudes, so that in a given volume of air more oxygen would be delivered to the engine, hence enabling the engine to produce more power.

When the "military power" phase of the mission had been completed, and the air group was approaching 20,000 feet of altitude, the air group commander gave the signal for all of the pilots to shift into "high blower" or supercharger mode to climb into the higher thin air above. As indicated, in high blower mode, the air is compressed to provide the same or nearly the same, amount of oxygen, which air at lower altitudes provides the engine. The aircraft had just gone into the supercharger mode when Ritter, as he often did, thought about the beautiful sight of so many aircraft in the brilliant blue sky and the special thrill of being part of such an activity. "If only my friends and family could share the experience of seeing the beauty and precision of this group of over one hundred airplanes in formation between the bright blue sky and the darker blue sea beneath," thought John Ritter. But, of course they could not — and what a shame. Before Ritter could finish his thought, the message "Mayday! Mayday!" broke the disciplined silence and was followed by a calm but piercing, "This is Paperweight 418; I have engine failure." The "Mayday!" got everyone's attention; but no one's attention was more riveted than Ritter's, because Paperweight 418 was being flown by Winston Gray, Ritter's best friend, who, along with his wife, had an adjoining apartment in the same house in Norfolk, Virginia. There were so many things the couples did together; and as Winston's Skyraider started to descend from about twenty-seven thousand feet where the air group had been operating, Ritter found himself thinking about all sorts of things: Uppermost was the thought that, although he had seen and heard about many things happening to many pilots, this situation brought home more clearly than all the others, that things can happen to those close to you and the ever-present possibility that you yourself are not immune to trouble. Would Winston be able to restart his engine on the way down? Had he run a droppable tank dry? Had the baffles in the fuel tank that kept the fuel from sloshing around come loose and covered a fuel intake? After all, the baffles had been a problem earlier that the technical representatives had tried to resolve. Was it a carburetor problem? No matter what Ritter considered as a possible source of the problem, he considered it very unlikely that a restart could be made. Ritter knew Winston would have to consider his options: either to bail out or to ride it all the way down to a water landing. Ritter knew also that Winston Gray was a very good pilot who would make the right decision. Bailing out presented its own set of problems as did landing in the water. But the Skyraider was a aircraft which pilots had landed in the water before with good results. With lots of altitude and a huge expanse of water which could be his landing runway, Winston could easily line up into the wind and, barring unexpectedly rough seas, would have a good chance of contending with the waves or swells in this apparently calm sea. He would have plenty of time to think about and do all the things necessary to prepare for a good water landing. Winston radioed that he would ride it down, and the air group commander immediately dispatched Paperweight 404 to escort him all the way down and report on his status. Ritter really wanted to be the escort pilot following Winston down because they were such good friends; but the air group commander had made his decision and Ritter respected his authority and logic, so that was that. Even if he could not observe Winston all the way down, he felt confident that Winston would handle his emergency well: he would make a good landing into the wind, hitting the water as softly as possible; he would get out of and clear of the aircraft, inflate his Mae West life preserver, and inflate and get into his small rubber raft, which had been packed in the seat of his parachute harness. But Ritter could not be certain that what he had envisaged would turn out that way. Aviation was like that. Sometimes conditions turned to worms; and what appeared to be relatively good percentage-wise turned horrendous. He had seen strange, even fluky, things happen in his brief four years of flying. So Ritter and everyone else in the formation could not be sure things could not become worse. What if additional problems developed? If his engine failure had been due to a fuel problem, could fuel be leaking with the potential of causing a fire?

Thinking about these things, Ritter tried to be positive and optimistic. "But what if this was one of those occasions when things went wrong?" He tried to put out of his mind the thought that, if the worst happened, he would have to escort Winston's body home and, along with others, try to comfort his wife and family. How silly even to think about such things — everything would be all right, and he would soon hear that Paperweight 418 had made a satisfactory water landing and that everything was okay. So, after what seemed an eternity, he heard the radio call to the air group commander: "Paperweight 00, this is Paperweight 404. Paperweight 418 is in the water and exiting his aircraft safely." This was followed shortly later by, "Paperweight 00, this is Paperweight 404. Paperweight 418 is in his raft and signaling that he is okay." A destroyer escorting the *Coral Sea* was dispatched to pick up Winston and had him safely back aboard the *Coral Sea* in an hour and a half. The air group commander was relieved, as was everyone else; but to John Ritter, Winston Gray's safe return was special because you could never, ever tell how things would turn out, even when conditions seemed ideal. He hated to think about what might have been otherwise and all that that would have entailed. Although conditions for making a landing in the sea could have been much worse, Ensign Gray deserves much credit for keeping his wits about him and for following the recommended procedures for handling his emergency — he remained calm and did things right.

There would be plenty of time later to talk about Winston's experience and what might have caused the engine failure. Several things came readily to mind. But for now, Winston Gray was safely back aboard his carrier. Winston had been calm, cool, and competent in this emergency, putting into action the emergency procedures the Navy had taught him. He had always exhibited good professional qualities; in addition, he had been completely reliable and extremely conscientious. Ritter knew that Winston would be on the next morning's flight schedule as expected and also as Winston wished — and that he would continue to exhibit the same professionalism. The couples would talk about Ritter's experience, perhaps even with a touch of humor. But in a quiet moment, Winston confided to Ritter that his approach to flying had changed: he would continue as he had always done, but the thrill of flying military aircraft was gone — not to be regained. He had reached that point in his aviation career that all professional military pilots attain after an early, and sometimes lengthy initiation: flying was not fun, but just plain hard work best done in a serious professional manner. He knew, as every military pilot knew, that no matter how skillfully you had handled a challenging flight today, you would have to go out and do it again and again and again. Yes, there were personal satisfactions of a job well done, like any other job with performance-requiring skill; but one bad performance out of many good ones seems to linger forever.

Weather Plays a Cruel Trick on the VIP's

Weather is, in all flying, an important factor, never to be ignored. Aircraft can rendezvous in clear weather or below or above cloud layers. They can rendezvous at night, the red exhaust of jet engines remaining visible even at relatively great distances. But it is practically impossible to rendezvous in the clouds. It is possible, if not dangerous and ill-advised, for a section or a small division of three, or possibly four aircraft to fly in formation in clouds. Encountering a cloud layer unexpectedly is usually trouble for a formation, resulting usually, but not always in the scattering of the aircraft in all directions. Small formations or a section could usually stay together without too much difficulty in the clouds, provided they stay in close formation. If one aircraft loses sight of the others in the clouds, trying to reform becomes extremely hazardous. Large formations would probably not encounter cloud conditions since special planning is required. But small groups occasionally do, sometimes resulting in tragic midair collisions. One example of jet aircraft in formation letting down thousands of feet of altitude through thick clouds at night was described in Part One.

One very unusual incident involving both a large-scale formation and fog is notable for its rather humorous side, perhaps not to all but at least to some, since it had a safe and happy, if very embarrassing resolution. During the latter part of a week and into the weekend, most of the aircraft from an entire air group were hoisted aboard a Midway Class aircraft carrier, which was tied up at Pier Seven at the Norfolk Naval Base. The pilots reported aboard as necessary on Sunday so that the carrier could make a very early departure on Monday morning and proceed to the operating area east of Cape Henry. Also on board, and the reason for the exercise, was a group of some two hundred VIP's from the Washington, DC, area. Not known to most pilots was exactly who these people were and what areas of government and/or industry they represented. It was assumed, probably with a fair degree of accuracy, that most of the guests were congressmen, accompanied by officials of the Defense and Navy Departments and important businessmen. The plan which called for the hoisting aboard of all the air group's aircraft, a very time-consuming process, would enable the VIP's to observe the morning launching of the entire air group. The VIP's would then observe some precision bombing, rocketing, and strafing runs, and then afterwards see the "recovery" of all one hundred twenty aircraft back aboard ship.

Whatever the air group commander and individual squadron commanding officers — good, competent people all — discussed and reviewed prior to the launch Monday morning, it is certain that weather was included, because weather is always something that has to be considered — and all the more so when huge formation flights are being planned. Besides, it was a crucial factor in the success of this mission. Whether the possibility of fog, the lowering of the temperature to the dew point temperature, was a part of the weather briefing we never heard. Even if the spread between the temperature and the dew point was less than three degrees, a somewhat risky area, the clear skies and the bright sunshine would augur well for an increase in the spread and the reduced risk of fog. Any patches of fog would likely burn off in short order.

Individual squadron briefings were conducted in the separate squadron ready rooms; and soon the sound of "Pilots man your aircraft!" was heard over the bullhorn. The takeoffs were accomplished in timely order; and before long, the rendezvous of the entire air group, more that one hundred aircraft, was accomplished. The pilots expected that they would soon be broken up into squadron units to begin their bombing, rocket, and strafing runs on the "sleds" or targets being towed at a safe distance behind the aircraft carrier. Intent on keeping their places in the huge formation, it required varying amounts of time for each of the pilots to notice an unexpected development — the slow growth of a white blanket of clouds, actually fog, below. Nothing apparently unusual here because formations of aircraft are often over, under, or between cloud layers. Circling in a huge race track pattern, the pilots soon observed that the sea below was completely obscured by the fog bank. Then, only gradually because each pilot had to concentrate on maintaining his close position in the formation, each pilot became aware of a strange sight, something which no one in this group had ever seen before: the gray antennae attached to the carrier's island superstructure came into sight, sticking up above the white fog and presenting a very strange, ghostly sight. The carrier's flight deck was about sixty-five feet above the water; and the island superstructure, which contained the Flight Operations and staff, and also the "bridge", or captain's space, was another forty feet high. On top, were antennae for various radars and other pieces of electronic equipment. All together, the tops of the antennae might be as high as one hundred fifty to one hundred eighty feet or more above the water. Though few if any of the pilots seemed to think about it at the time, the aircraft carrier's commanding officer and his staff had their own problems to deal with concerning what decisions they could and should make about the fog. There was no danger of the carrier's colliding with other shipping because the carrier had excellent radar, but the C.O. had to think about the circumstances and how he could continue the demonstrations planned for this day. The ship's search for a clear area in which to operate became crucial.

The air group continued its race track holding pattern, coming alongside the starboard side of the antennae every few minutes. The pilots thought that at any minute the dense fog would be dissipated gradually by the bright sun as it often had and that in short order the show could begin; or, at the very least, that the carrier would soon move out from under this particular fog bank and into a clear area where operations could begin. Perhaps it was just an area of water, cooler than other areas, which had cooled the air above to its dew point where it condensed into fog. If so, would not the ship be able to move back or farther on to a spot where the water was less cool? After all, the pilots had plenty of fuel, the carrier was not far off the coast of Virginia, and time did not seem to be a problem. But before they were prepared for it, the pilots received the surprising word that the air group would depart immediately to return to land at Breezy Point, the Norfolk Naval Air Station. The carrier, its crew, and the distinguished visitors were left in the fog, both literally and figuratively. What the VIP's thought of the operation the air group's pilots never learned. The VIP's had seen an impressive launch and rendezvous of over one hundred aircraft. A little later and while in the fog bank, they could hear the roar, perhaps even catch a glimpse, of the aircraft overhead and passing along the starboard side of the carrier, since visibility is usually much better vertically than horizontally — particularly with a bright sun overhead.

Whether the decision to return to base was wise the pilots likewise never learned. To the pilots, it seemed a rather hasty decision under the circumstances; but predicting the dissipation time of fog is often very difficult, and they were never privy to the communications between the air group commander and the aircraft carrier, their transmissions being conducted on a separate radio frequency. If the decision was not wise, it was, nevertheless, decisive — and safe! No doubt, safety was the prime consideration. Any kind of an accident or tragic consequence in the existing weather conditions would have been unacceptable. It would have been grossly negligent. Though curious, and after landing back at Naval Air Station Norfolk, the pilots did not feel that questioning the air group commander, or even the squadron commander, was a prudent course of action. You just didn't do that. If that seems strange, consider also the fact that the "skipper's" judgment rarely seemed questionable; and the discipline that produced such good flying skills and procedures depended upon a confidence, almost always a well-placed confidence, in the leadership of experienced senior officers — and not in trying to second guess them. And so on this day, the pilots also remained in a fog of a different kind concerning the day's events.

The only topic of conversation among the air group's pilots after that "interesting" flight dealt with the decision the aircraft carrier's aerology department made in determining that the difference between the air's temperature and the air's dew point would not drop enough for fog to form. They wondered what kind of hell the aerologists would receive for making a faulty judgment. How embarrassed would the carrier's commanding officer become? How would he react to the senior aerologist? Suppose the air group had been launched in mid-ocean and could not get back aboard. Certainly, there would have been much more caution in mid-ocean. However the decision of the aerologists was made, most pilots did not seem to fault him. After all, when they awoke, had breakfast, and launched from the aircraft carrier, the weather was beautiful. Moreover, normally on such a sunny day, the air is heated by the sun, raising its temperature and increasing the spread between the temperature and the dew point. On the VIP's day out at sea, that just didn't happen!

The captain had made his decision to launch the air group on the best information he had, hoping and expecting — perhaps with his fingers crossed — that the weather would allow "the show to go on." Many people would argue that "air shows" are just bad news. Many times, something goes wrong. Reports of such events and incidents appear in the press, events with tragic consequences. A relatively harmless incident at the Marine base in Quantico, Virginia, comes to mind. Various Navy and Marine squadrons were demonstrating their strafing and bombing skills in front of a grandstand full of VIP's from Washington, D.C. Many demonstrations went smoothly and as intended, until a Marine pilot, flying the F3D Skyknight, and his radar operator were making low altitude radar bombing runs. The radar would lock on the intended target and the bomb would be dropped at the appropriate spot. But the fire control radars had what was called a "range gate," which would not only lock on a target but also allow the operator to drop that particular radar target and search farther ahead. In some cases, the radar might skip over a smaller target to lock on a larger target. In this case, some metal buses, like school buses, which had brought the VIP's to the grandstand, were parked straight ahead of the intended target and along the planned flight path of the Marine aircraft. These buses provided a much bigger target, which the Marine's radar locked on. Whether the buses were damaged was not known, since the bomb was miniature in size and the pilot could have intervened. At any rate, damage or not, the story became widely known.

Swept-Wing Jets Go Aboard the Carrier: A Last-minute Change

Preparations were underway to have a group of eight pilots flying F9F-6 Cougars, new jet fighters with wings swept back thirty-five degrees go aboard the *USS TARAWA* (CV-40), a straight-deck Essex class carrier of about 27,000 tons, for pilot carrier qualifications in the newly assigned aircraft. In preparation for landing aboard ship, the pilots would first have to make simulated landings or "Field Carrier Landing Practice" (FCLP) at their home base. The only suitable times for field carrier landing practice at home base were well before 0800 or after 1630, times when most of the routine flying schedules of the three large composite squadrons assigned to the Naval Air Station were not being carried out. For the pilots practicing field carrier work, that meant showing up at 0630 AM and remaining till after 1830. That made a long day, but the effort was considered necessary; and there was no compensatory time granted, nor were any pilot complaints aired. Despite having the gumption to fly demanding aircraft in demanding conditions, they wouldn't have dared question the skipper's orders just because they were putting in long days. Besides, a war was on, and any balking at such requirements would be decidedly out of place.

The practice procedure for field carrier landing practice involved taking off with adequate spacing between aircraft, climbing to 200 feet altitude, turning left downwind, and then, when opposite the landing end of the runway, making a turn toward the runway while descending to just above the tall pine trees and slowing the aircraft down to a speed of 115 knots (just over 132 miles per hour). Since there was no need to rendezvous, the pilots were responsible for creating the required distance or interval between aircraft so that the practice could move on in an orderly fashion. The speed of

115 knots was not a comfortable speed for this swept-wing aircraft because it seemed the aircraft was wobbly and less controllable, a characteristic prevalent in swept wing aircraft. But that was the procedure; and the eight pilots practiced, practiced, and practiced at 115 knots. Now, with all the F9F pilots having made many simulated carrier approaches and landings, they were ready to fly aboard the carrier, a straight-deck aircraft carrier of the ESSEX class.

For a pilot making an approach to land on an aircraft carrier, the proper and precise airspeed is crucial. If the aircraft is too slow, and particularly in the approach turn to the carrier, it can reach its stall speed where it does not produce enough lift to stay in the air — and at such a low altitude, there is no way to keep from crashing into the water. If the airspeed is too fast and the pilot is not "waved off" to make another approach, it will overshoot the intended spot on the carrier deck and slam into the restraining cables — another crash but usually of less serious consequences. Either situation is bad. So the airspeed indicator is the most important of many important instruments during the carrier landing procedure.

At the pilot briefing for the flight out to the carrier and subsequent landings, a surprise announcement was made. As a result of some very recent experiences of other pilots from other units who were doing carrier qualifications in the Cougar and other jet aircraft, it was determined that 115 knots was too slow a speed to land aboard the carrier. The new speed was to be 125 knots (nearly 144 miles per hour). The pilots received this news with mixed feelings: The higher speed would reduce the slow-speed wobbling and make the aircraft feel more controllable, further removing it from the stalling speed; but, for a straight-deck carrier of the ESSEX class, the added speed would produce a tendency to touch down or land farther up the deck and closer to the barricades, a device of stretched cables running athwart the deck which captures aircraft falling to engage its tail hook in one of the arresting wires. Although the increase to 125 knots at first seemed strange, the decision was based on sound reasoning. At 115 knots, the aircraft was flown in a higher nose-up position than would be required at 125 knots. And pilots, who had flown aboard before this group of eight, had to make their approaches in this higher nose-up attitude, much as propeller-driven aircraft had. While flying the aircraft in this higher, slower, nose-up attitude was really not a problem, another, more serious problem resulted: In this higher nose-up attitude, a situation was created in which the aircraft's tail hook extended lower and could catch an arresting wire before the main wheels were on the deck; the result was that when the hook caught the wire, the nose was whipped downward so that all of the aircraft's weight was forcefully thrown onto the nosewheel and nosewheel strut before the main wheels touched the deck of the carrier. The stress on the nosewheel and supporting strut was such that almost always the cast aluminum nosewheel strut shattered. Indeed, there had been enough of these accidents to prove the point; and even with the increased speed of 125 knots, some of the landings were such that the nosewheel was whipped to the deck with the same results. A report, not corroborated by a second source, mentioned that this same type of problem existed when the swept-wing F7U Cutlass landed aboard in a higher than desired nose attitude. However, the same source reported that, when the nosewheel strut shattered on being whipped down to the deck, the pilot whose seat was positioned almost directly above the nosewheel and strut received back injuries, sometimes including fractured vertebrae.

But for this group of eight pilots going aboard for the first time in Cougars, there was a concomitant specification: that the aircraft carrier would insure that the combination of its speed into the wind and the actual wind force itself would equal forty knots (46 miles per hour) and even hopefully to fifty knots (57 miles per hour). With the wind force at forty knots, the pilot would have a relative speed as he crossed the stern or fantail of the carrier of about 100 miles an hour; in effect, this meant the deck would be longer, allowing more room for the higher speed now required. For the pilots, that was a condition "devoutly to be wished," because they knew that sometimes carriers encounter calm or nearly calm situations; and even at full speed in calm winds, an ESSEX class carrier would not be likely to sustain speed above 25 knots, resulting in aircraft relative approach speed of 115 miles per hour, a significant and perhaps dangerous speed.

The change in the aircraft's approach speed to 125 knots, made an annoying situation they had been contending with worse. For some unknown reason, the airspeed indicator, an instrument on the flight panel, was located at the lower left side of the instrument panel at the lowest point of the panel. For any landing approach, the pilot must keep both looking out at the field or carrier and, at the same time, keep referring to his airspeed indicator. The airspeed indicator should be located near the top left part of the panel. The lower the location, the more the pilot must keep bobbing his head up and down from the aircraft carrier to the airspeed indicator. Even in normal approaches, some bobbing is necessary, but the location of the airspeed indicator exaggerated it. Most Navy landings involve left or slightly left turns to the field or carrier, dictating that the essential flight instruments be convenient to the line of sight to the field or carrier. Not only should the Navy flight instruments be generally left of center or in the center, but they should also be easy to spot as the pilot shifts his sight from the carrier to the instruments and back in a continuous process till touchdown. But in all or nearly all of the Cougars in the squadron, the airspeed indicators were located at the bottom left section of the instrument panel, a strange and poor location considering the importance of airspeed.

The pronounced head and eye shifting required by the location of the airspeed indicator was annoying, but it was not the worst thing. It was made worse by the fact that all of the instruments were recessed by about three eighths of an inch by the lighting panel which is superimposed over the entire instrument panel. The result was that the top fifth of the airspeed indicator was covered or blocked from view, no speeds in this fifth being visible at all unless the pilot squeezed himself way down in his seat. This scrunching of the body meant that the pilot spent more precious time with his head down in the cockpit, rather that looking out at the carrier and positioning his aircraft to line up properly with the center line of the aircraft carrier. This situation, laughable as it may seem were it not so serious, was compounded by one more problem resulting from the location of the instrument: The top fifth of the airspeed indicator, which was not visible, contained the critical approach speeds, including of course, the necessary numerals <u>125</u> for 125 knots, making 125 the most difficult of all the numbers to see.

How each pilot coped with the situation is not known, nor is it known how many complained. But no one seemed even to talk about it or know whether any accidents had been the direct or indirect cause of the instrument location. Military pilots, unlike airline pilots can't scream "safety of flight!" to get something done. On the contrary, Navy pilots traditionally have felt that they should be able to cope and therefore should not complain. If they felt they should not complain, it does not mean that they did nothing about it. The poor location of the airspeed indicator was one thing, a thing that could not be remedied overnight. Not being able to see the 125 knots figure on the airspeed indicator because of the instrument location was another thing, a thing about which some pilots felt they could do something. Even if the pilots could not see the figures or the arrow point at the top of the needle, they could see a good part of the needle. A couple of pilots, using grease or china marking pencils, drew a straight black line on the instrument's glass cover extending from the 125 knot figure down to the center of the dial. Since the needle of the airspeed indicator rotated from the center point of the dial, the pilot could easily see the point of the needle disappearing behind the marking pencil line on the dial. If he was getting slow, the needle would be slightly to the left of the line; if too fast, the needle would show slightly to the right. While nothing could make the situation perfect short of moving the airspeed indicator to a better location, a simple grease pencil did wonders. Some pilots wondered, after having had this experience, whether it might not be a bad idea to have a movable needle installed which could be set or fixed in a position to indicate the precise airspeed to be maintained. Looking at the grease pencil mark and believing that the airspeed indicator needle was behind it did require an act of faith to be sure you were flying at exactly 125 knots when you could not actually see the figure, but those lines were drawn very carefully and they really did the trick. Is it not strange that, in sophisticated and well-engineered aircraft, so much depended on so crude a tool as a marking pencil to compensate for faulty design. There was another twist to this day's flights: at no time did the wind across the deck -that is, the actual wind plus the aircraft carrier's speed — exceed 25 knots. The pilots were not aware of it till later, and no one experienced any unusual difficulty — though, no doubt, the landing signal officer was stressed, having to give the "cut" signal a bit sooner when the aircraft was farther from the fantail of the carrier than he would have preferred had the wind been stronger.

The reader can perhaps sense that, at the time of these Cougar pilot carrier qualifications, aircraft carrier operations were in a state of change as a result of the requirements of jet aircraft. In another section, there is some explanation of the jet engine's restricted rate of acceleration when the throttle was advanced suddenly to the full position. The jet engines contained a fuel metering device which fed fuel to the engine as rapidly as it safely could. Without this device, the sudden application of fuel via the throttle would place so much fuel into the engine that the engine would just quit in what is called a "rich out," the opposite of a "flame out," which occurs with fuel starvation. It should be noted that there is an emergency throttle system, or fuel-flow system, for use should there be a problem with the normal system. Shifting to the emergency system gives the pilot a useful but dangerous manner of throttle operation, one that still requires careful operation to avoid rich out and exceeding the RPM limitations of the engine.

The jet engine delay required by the fuel metering device when the throttle is suddenly advanced to full throttle could take up to eight seconds for the engine to accelerate to full power. For jet aircraft landing on an straight-deck aircraft carrier, wave-offs could and did cause some severe problems due to the slow engine acceleration, as one pilot's experience indicates. A very capable and experienced pilot was making a satisfactory approach to a straight-deck aircraft carrier in an F2H-3 Banshee jet, a model quite a different aircraft from the F2H-2 Banshee, and at the moment he expected a "cut" signal to land aboard, he may have reduced his throttle setting just a tad, only to receive an unexpected "wave-off," signaling him to go around for another approach. He applied full throttle to the two jet engines, having no choice but to continue up the center line of the carrier for fear a turn might exacerbate his situation and to hope he could maintain and perhaps even increase his altitude to insure his clearing the barricade or barrier cables stretched across the deck. The slow engine acceleration kept him from climbing sufficiently to clear the barrier, and his nosewheel caught the top cable of the barrier and was snapped off. His aircraft was then slammed to the deck forward of the barrier, where some aircraft had been parked. His plane bumped a parked aircraft, sending it teetering on the port edge of the deck, while the wing of another parked aircraft scraped the edge of his cockpit. His aircraft, now in flames, slid toward the bow of the aircraft carrier. Sliding forward, his aircraft turned slightly left as it approached the bow of the ship, stopping just short of the bow. The wind over the deck blew the flames aft, and the pilot climbed out on the bow side free of the flames and onto the few feet of remaining deck space between his aircraft and the bow. He was, quite naturally, shaken, but otherwise not really hurt. It will come as no surprise that, given the nature of Naval Aviation and its pilots, he was flying again the next day, never again to experience an accident of any kind.

* * * *

The special requirements of operating jet aircraft aboard aircraft carriers became quickly obvious, leading to the first aircraft carrier to have an angled deck. The *USS Antietam* (CV-36, CVA-36) had been modified to have an angled or canted deck in the early 1950's. Subsequent developments included the first aircraft carrier designed with an angled deck to provide for the requirements of jet aircraft. It was the 60,000 ton, 1036 foot long *USS Forrestal* (CVA-59), and was launched in 1954. It was named after the widely respected James V. Forrestal, the first Secretary of Defense, whom President Truman appointed on September 17, 1947. Some quotes from the Naval Aviation News (November-December 1993) about the *Forrestal* relate to the conditions described in this chapter.

"One major problem with early jets that had an enormous impact on safety was very quickly identified. Early jet aircraft suffered from slow engine response to throttle inputs by the pilot. Whereas piston engines responded rapidly to requests for full power, the jet engines required time to 'spool up.' Rapid throttle response was mandatory for aircraft landing on carriers at sea. Increased danger was also experienced because of the straight deck design of the Essex and Midway classes of carriers. An aircraft could not go around (for another approach) if it failed to engage an arresting wire; and despite the series of barricades meant to protect the parked aircraft forward on the flight deck, collisions were common.

"Early in her construction, *Forrestal's* design was altered to accommodate new technology that would eventually solve the safety problems experienced when operating jet aircraft from carriers. First, it was decided to incorporate an angled deck into her design. This meant that aircraft on final approach would have an unobstructed path, enabling them to either touch and go — that is to bolter — or to trap on deck. Second, steam catapults were installed in place of the originally planned powder-charged ones... Interestingly, the three innovations that enabled safe operations aboard carriers — the angled deck, the steam catapult, and the mirror landing system — were all originated by the British." It is worth noting that the Navy had used hydraulic fluid catapults until leaking hydraulic fluid caught fire aboard the *USS Bennington* (CV-20) in May of 1954, resulting in the death of many crew members, some still sleeping in their bunks.

Part Six: Instrument Flight

Instrument Flight — Essential to Life

An earlier section described some tense moments that jet pilots experience with the rapid rates with which military jet aircraft consume fuel. In particular, jet fighter aircraft, which may have to change altitudes frequently in pursuit of other aircraft that are taking evasive action, consume fuel both in the high power settings of pursuit and the high rates of fuel consumption when pursuit takes the aircraft into lower, less fuel-efficient altitudes. That section also emphasized that the nature of jet aircraft with their fuel-consuming engines requires that they climb to high altitudes (40,000 feet or thereabouts) where their engines can operate most efficiently and economically. But climbing to the high altitudes often means that the pilot must climb through clouds or layers of clouds; and even if he can climb to his high operating level free of clouds, he must be prepared to have to descend through different conditions — that is, through cloud layers or solid clouds. To operate in these conditions, the pilot must be proficient in instrument flying procedures; in other words, he must develop the skills necessary to fly in the clouds by reference to cockpit instruments alone, not by relying on outside visual references — because there just aren't any outside visual references — and he must also develop the skills necessary to make landing approaches to the runway in challenging conditions of very low clouds and very poor visibility, not to mention icing during the descent and then landing on snow-covered or ice-covered runways. While it may seem hard for non-aviators to believe, when a pilot is completely immersed in the clouds, he really cannot tell which way is up or down, whether he is turning left or right — unless that is, he refers to and trusts what the instruments in his cockpit tell him. However, without instruments he may sense whether he is gaining or losing altitude and whether his speed is increasing or decreasing. The sense of balance becomes unreliable because the sensors in the middle ear, which consist of upright hairs, move from side to side rather than remaining upright, giving false impressions. One need only think of a person who spins quickly and then stops, only to find himself dizzy; he recovers quickly because his visual references and the now stabilizing ear sensors provide the needed recovery. In flight in clear skies, any dizziness is overwhelmed by visual references; but in flying in clouds, those references don't exist, so the pilot needs to rely on the cockpit's instruments and not his faulty sense of balance. Indeed, for the fledgling pilot who trusts his senses more than his flight instruments, there can be tragic consequences; for nonbelieving pilots can become "spatially disoriented" and can, and have, paid the ultimate price for their lack of faith.

In Naval Aviation flight training, and for sure in other flight training programs as well, fledgling pilots are taught over and over again to trust their instruments. But when a pilot becomes disoriented and confused, he can think that the instruments are not providing reliable information and that his senses are more reliable. If he persists in that thinking while he is in the clouds, it is conceivable that his aircraft could be in such an attitude when he emerges from the clouds that he might not have enough altitude to level the wings and pull out of his descent before striking the ground. But student pilots in the military are instructed that they must learn to trust their flight instruments; and fortunately such incidents are extremely rare.

Four tragic civilian accidents, occurring over the past couple of years, underscore the importance of instruments in flying in challenging weather conditions. The FAA determined that John Kennedy's accident off Cape Cod was caused by spatial disorientation. Soon afterward, a pilot flying in New Jersey, who was supposedly experienced in instrument flight radioed that he was experiencing "gyro problems" (Gyroscopes drive several instruments, the most important of which provides an artificial horizon in the cockpit, a horizon which remains true to the actual horizon outside the cockpit.); shortly thereafter, he spun out of the clouds and was killed. And the third pilot, "supposedly instrument qualified," spun out of the clouds near the Coatesville, PA, airport. In all three cases, there was a strong consensus that spatial disorientation was involved, although it is unknown whether the FAA has made a formal finding in all cases. And most recently, there was the tragic death of Missouri Governor Mel Carnahan, his son, and an advisor in an aircraft accident, reports of which from the ground seemed to indicate spatial disorientation. One news article reported that the Carnahan family was sure that there was a failure of gyroscopic instruments. But the tragedy in all of these incidents is that, most likely, they were avoidable.

So, along with all that is required to fly the aircraft, takeoffs, landings, tactics, knowledge of systems, etcetera, the pilot must develop skills in instrument flying so that they become second nature to him, skills which he does not have to think about for a moment but which he automatically uses as part of any flight. For most pilots, basic instrument flying skills are learned rather quickly and then honed first with practice and then in use in bad weather. A few letdowns through thick clouds and approaches to landings in poor weather conditions help further to boost the pilot's confidence. And while all approaches to landings in bad weather conditions require flying or aircraft handling skill, careful attention to detail, and knowledge of letdown procedures for a particular landing site, the pilot has confidence that, if he does things right, he will make a satisfactory approach to a landing. Inherent in all of this is the pilot's knowing when the weather conditions are too bad or marginal for him to attempt an approach; likewise, he must know that, when he has made a poor approach, he must add power to his engine or engines, climb to a specified altitude, and execute a missed approach procedure, which will place him in a position to make another, and hopefully better, second approach.

Before describing how pilots acquire instrument flight skills and the flight instruments they will need to use, it would be advisable to mention a few very simple principles: the first is the "lift" provided by the wing in level versus non-level flight; the second on basic flight instruments will follow just a bit later in this chapter since lift and instruments are crucially intertwined in instrument flight, although they are also important in non-instrument flight.

Of course, this brief explanation covers but a small part of the aerodynamics essential to flying and to the pilot's background. But it will help to make clear those areas in which problems can arise and in which there are serious consequences. To emphasize the serious nature of understanding lift, some brief reference to a real incident is helpful. Along with believing

Instrument Flight 223

SIMPLE PRINCIPLES OF LIFT

Aircraft in level flight remain in the air and in level flight because the lift on the horizontal wings balances the weight of the aircraft. Referring to Figures A and B, it is easy to see that the lift of Figure A exactly balances the aircraft's weight. Consider what happens as soon as the pilot makes a turn, a left turn with 22 $\frac{1}{2}$ degrees of bank: the lift in Figure B, which is perpendicular to the wings, has two components, a vertical component now only three fourths of that when the aircraft was not turning and a component of one fourth, which is now directed in the direction of the turn. Since there is less lift when the aircraft is in a turn, the pilot must do one or the other of two things, often both, if he wishes to remain in level flight — that is, to maintain the same altitude: he must ease the nose of the aircraft up to get more lift from the wings, or he must add power to increase his speed to get more lift. In many cases, doing both is preferred; but situations dictate what is to be done.

Figure A   Figure B

Consider a turn with a bank of 45 degrees, Figure C. In such a turn, only one half of the lift is directed in the vertical direction, thereby requiring the pilot to act more vigorously if he is to remain in level flight. And consider Figure D in which the aircraft is in a 90 degree bank. In such a turn, there is no vertical lift component. So no matter whether the pilot raises his nose or adds full power, or both, he cannot produce any vertical lift. The result, of course, is that he cannot remain in level flight. If he wishes to keep from losing altitude, he

> **Simple Principles of Lift (Continued)**
>
> must move toward leveling his wings. In some big, heavy aircraft, considerable altitude will be lost before the pilot gets his wings level again.
>
>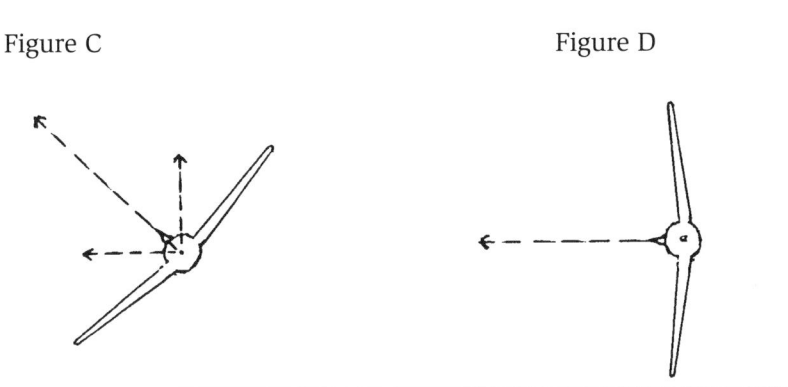
>
> Figure C Figure D

the instruments is another essential belief involving what to do in case of high G (gravity-induced) blackouts when blood drains from the pilot's head. Often caused by dogfights, these blackouts occur much less frequently today because pilots wear blackout suits which, by applying pressure at certain points, inhibit the blood's draining from the head. There is an attendant problem that goes with blackout training: Pilots who disregard the preflight training or who think that they know better can get into life-threatening situations. Most pilots take their training very seriously; but occasionally some student pilot like Rocky Levine seems to either ignore the training or become confused, perhaps even panicky in the blackout condition he finds himself in. Unless the pilot acts as he was taught, his condition, known as the grave yard spiral can prove fatal. But more about Rocky in a later section. While not an instrument flight problem, it is a confidence problem — believing what others have learned and passed on — and its relation to instrument flight is that just as one must believe his instruments, he should accept the belief of what to do in a blackout.

In both types of situations, lift forces are important, and they are related to wing position; that is that the lift forces which keep the aircraft in the air are perpendicular to the aircraft's wings. In tight turns, either as a result of

The Artificial Horizon
(Also known as the Gyro Horizon and the Attitude Indicator)

By means of a gyroscope, the Artificial Horizon instrument aligns its horizon with the actual or real horizon outside the aircraft. That horizon is maintained regardless of the aircraft's actual nose or wing positions. The miniature aircraft in the instrument is fixed in its position, thereby indicating the aircraft's attitude relative to the outside horizon, reflecting any changes in the aircraft's nose or wing position — or both. In other words, the pilot flying in the clouds depends on the information on the Artificial Horizon.

Three examples are provided. The first, center figure, indicates the aircraft has its nose on the horizon and its wings level with the horizon. The second, left, shows the aircraft in a descending left turn, while the third, right, shows a climbing right turn. The instrument markings indicate the degree of bank: 22 ½, 45, 67 ½, and 90 degrees.

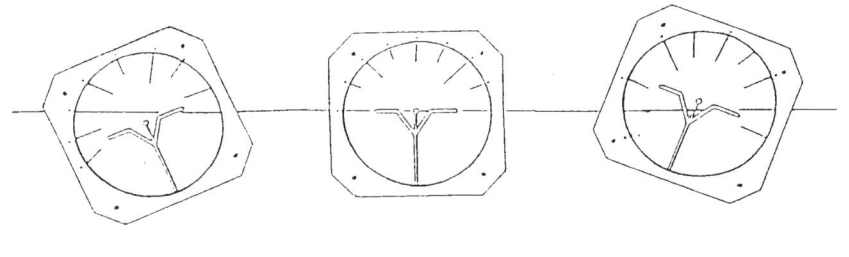

being blacked-out or disoriented, the lift is reduced so that it is insufficient to maintain level flight. The only solution is to level the wings.

In developing his instrument flying skills the pilot need not only know basic principles of lift, but also essential instruments. Without going into too much detail, there are a few instruments he must learn to use. Basically, there is a gyroscopic instrument known as an Artificial Horizon. (See the accompanying drawing of the Artificial Horizon.) A gyroscope, usually powered by a vacuum pump connected by an accessary connection to the engine, provides the means of presenting the instrument's artificial horizon parallel to the ground, while a small figure in the shape of an airplane stays parallel to the aircraft's wings. All the pilot has to do is keep the figure's wings level with the artificial horizon and his nose on the horizon to maintain level flight. If he wishes to turn to a certain heading, by reference to the instrument, he banks the aircraft's wings to accomplish a certain rate of turn, which he maintains until he reaches that heading. If he wishes to climb or letdown, he adjusts his engine power setting, at the same time adjusting his nose position to a point above or below the instrument's artificial horizon. Other instruments help the pilot refine his artificial horizon work. Mentioned by inference was a compass to indicate heading. Other important and reliable instruments include a gyro compass and a Doppler or radar altimeter. But most importantly, there is a group of instruments not dependent on the aircraft's electrical or other vacuum power. One crucial set of instruments in this group depends on outside pressure to provide airspeed, altitude, and rate-of-climb (or descent) information. This information comes from what is called a Pitot tube, a cylindrical pipe-like tube often slung under the aircraft's wing. Ram or slipstream air caused by the aircraft's motion through the air mass entering a forward-facing opening provides airspeed information, while small holes on the bottom or side of the tube allow static air to the tube to provide very precise and sensitive altitude and rate-of-climb information. Another crucial instrument, likewise needing no electrical power, is the Turn-and-Bank Indicator. It consists of a vertical needle which pivots around a pin at the bottom of the needle somewhat like and upside-down pendulum, which remains perfectly upright only when the aircraft is not turning. A delicate needle, it requires skill to keep the needle perfectly upright — or if deliberately turning, to keep the needle in a steady position to maintain the proper rate of turn. In the same instrument, there is a fluid-filled glass, much like the glass in a carpenter's level, with a freely moving ball. Used in conjunction with the Turn Indictor, the ball's position in the glass indicates whether a desired turn is a coordinated turn (ball in the center of the glass) or whether the plane is in what is called a skid or a slip (ball off to one side or the other). A pilot desiring to make a timed turn on instruments will want to make a coordinated turn so that he will know more precisely his true rate of turn. Scientific developments in aviation and in aviation instruments seem to keep pace with scientific developments in other areas, so it is not surprising that improved instruments have come upon the scene over the years. One of the first was a gyroscopic

The Life Saver — The Turn and Bank Indicator

The simplest and most dependable attitude indicator, the Turn and Bank Indicator, needs no internal electrical or vacuum power. Nor does it, like the altimeter and air speed indicators, rely on ram air or static pressure. It is simply a mechanical device employing an inverted pendulum-like needle to indicate degree of bank and a ball floating in a liquid to indicate whether the turn is a smooth turn — that is, free of skid or slip. <u>But, its presentation does not indicate altitude or nose position, so it must be used with the altimeter</u>. However, in situations where the Artificial Horizon is disabled or its gyroscope has tumbled — and perhaps where pilot's spatial disorientation has set in — getting the wings level is crucial to survival.

Three examples are presented which relate to the illustrations on the Artificial Horizon's wing positions; but of course, they provide no nose position information. The first shows wings level; the second, lower left, shows a left turn; and the third, a right turn. The two showing turns are one needle width turns or twenty-two degrees of bank. So simple! Note: In flight, and as part of the aircraft, the left and right examples would actually be tilted as the Artificial Horizon left and right figures.

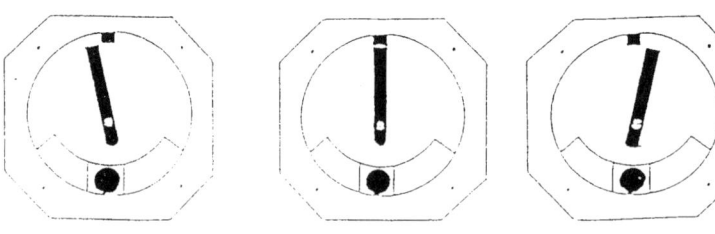

compass, which at first, at least, did not eliminate the magnetic compass but complemented the magnetic compass. It had the desirable feature of indicating the precise heading during the turn as opposed to the magnetic compass which had a "wobbly" effect until the aircraft was steady on a new course. Then, there was the "Doppler" altimeter, which provided actual height above the terrain as opposed to the pressure-sensing barometric altimeter, which presented the theoretical height of the aircraft based on the barometric pressure at different altitudes. Whereas the barometric altimeter always measured the altitude above sea level, it did not indicate the altitude above the terrain. The Doppler altimeter is sometimes called the radar altimeter because it bounces waves off the terrain below to measure the actual distance to the ground. At present, Doppler-based instruments are much in use with the weather bureau. Instruments based on gyroscopes like the artificial horizon have improved so much that they are almost tumble-free, eliminating the tendency to tumble out of control if a turn becomes excessive. Computers have made their way into aircraft cockpits. And very recently and still in a transition stage, cockpit instrument information is being displayed digitally in what is referred to as "glass cockpits." Information is displayed on screens much like computer screens. In larger military aircraft, as in commercial aircraft, with the luxury of two or more engines, there is designed redundancy of equipment, so that in the unlikely event of vacuum, electrical, or other failure, another system remains operable. The military aircraft has many dials, switches, panels, controls, etcetera, with multi-engine aircraft having hundreds. But no matter how many instruments and switches or how seemingly complicated the cockpit instrument array appears, and while all of the other indicators provide useful information or provide enhancements, the pilot depends on just a few primary instruments to fly his airplane — whether in clear air or immersed in dense, vision-obscuring clouds. Only two are presented here to emphasize the single most-used instrument regularly used in flying on instruments (The Artificial Horizon) and an instrument (The Turn and Bank Indicator) which can be used as a backup in emergencies or when the artificial horizon's gyro has "tumbled" or is otherwise unreliable.

It is necessary to recognize how important the Artificial Horizon is. One company advocating the use of a second or additional Artificial Horizon, an electrically driven gyroscope as opposed to a vacuum-powered instrument, uses a frightening quotation from *Flying Magazine* to stress the desirability of having a backup instrument: "The accident data base and university research seem to agree: 'If your attitude gyro (Artificial Horizon) quits while you're in thick soup, you're probably gonna die. So how come you don't have a backup?'" However desirable a backup may be, it might not be the solution to problems. An incident described herein shortly points out what can happen.

Despite the improved electrical generators and electrical systems which have provided greater reliability, there is still the possibility of failure, more likely of course in single-engined aircraft. What then? Depending on the nature of the failure, it is possible, though perhaps increasingly less likely today as equipment becomes more reliable, that the pilot will have to rely on those instruments which do not depend on electrical or other power sources. So, we get back to using those instruments like the turn-and-bank indicator, and those instruments (the air speed indicator, rate-of-climb indicator, and the altimeter) which operate through the Pitot tube to provide ram and static or barometric information to the those three gauges. In such conditions, it is likely that the pilot will also have engine instruments to indicate power settings. Consider then a situation in which a pilot flying in the clouds suddenly loses his gyro instruments for one reason or another and shortly thereafter encounters severe turbulence which upsets his level flight condition and "scrambles" the turn-and-bank indicator, altimeter, and rate-of-climb indicator. Aside from any navigational problems he may have as a result, his immediate concern is to get and keep his aircraft in an upright and stable condition. The pilot then is likely to have only the turn-and-bank indicator, the altimeter, the air speed indicator, and the rate-of-climb indicator available to him, the use of which, for obvious reasons, is called "Partial Panel" instrument flight:

Practice in the use of partial panel instruments is provided in the instrument flight training phase of the pilot's overall training, (all of which relate to the situation just described). Keeping in mind the recent tragic, weather-related accidents mentioned, put yourself in one of their situations. There will be three sets of experiences you as pilot will have to deal with on "partial panel:" first, is getting control of the aircraft which has just been lost by disorientation or jostled by turbulence to the extent that the wings are not level and the aircraft is in a nose down or nose up position; then later when the pilot realizes that he still has a safe, reliable, though more hard-to-manage condition, he realizes that he can make controlled climbs or descents with the wings level and with the adjustment of engine power; he also realizes that he can make coordinated turns at a controlled rate of turn. The last may include both climbing or descending at a controlled rate of descent, while at the same time making turns at a controlled rate of turn to get to a desired location. Although these maneuvers seem ridiculously simple, and in fact are, the very sensitive nature of these instruments makes these tasks a bit of a challenge — you get reliability but you pay for it in having to deal patiently with sensitive instruments. If you, the pilot, do everything in the proper manner, you will be able to handle both the conditions you find yourself in and the requirement to proceed under those conditions. If not, consider the consequences. What is the price you pay for not knowing or not having the patience or skill to operate on partial panel?

Pilots gets the instrument experience in "partial panel" flight that they need through several means. That experience can be gained through aircraft simulators, which are increasingly used in pilot training and provide practice for all phases of flight; and it can be obtained in actual flight where the student pilot's vision is shielded or blocked so that he cannot see outside of the cockpit, a condition commonly referred to as "under the hood." The instructor has the student "cage" or lock the Gyro Horizon, after which he makes a series of rather sharp turns left and right, pulls or pushes the controls (or stick) up and down, repositions the throttle, and then tells the student pilot, who now must fight vertigo, to take over and place the aircraft in wings-level, controlled, level flight. Getting the wings level first is of most importance to control attitude. In other than wings-level flight, it is possible that no amount of pulling the nose up will help at all, but only make things worse — perhaps disastrous. The pilot-in-training would first of all level his wings by use of the turn-and-bank indicator; then he would attempt to obtain level flight by reference to the rate-of-climb indicator and the altimeter, at the same time perhaps adjusting his engine power setting if that is necessary; then he would maintain level flight, again by use of both his altimeter and rate-of-climb indicator. All of this cannot be accomplished and maintained in an instant because of the sensitive nature of the partial panel instruments, particularly the turn and bank and rate-of climb indicators and to some extent the altimeter, so time and patience are required. In Naval Aviation instrument flight training, such exercises were called "Unusual Attitudes" and were experiences which provided lasting impressions. One student, so befuddled by the training could not help but confusedly refer to them as "peculiar positions," a term which, however descriptive, never failed to provide laughs when mentioned. After many such unusual attitudes on a flight, the instructor would have the student pilot go through the climbs, turns, and descents mentioned above on "partial panel." Practice in these procedures is not very exciting, and the unlikely possibility that partial panel flight will become reality make the practice of these procedures unappealing. Yet, the serious consequences of the neglect of such practice should override any tendency to negligence.

A June 6, 2002 newspaper article in the *Philadelphia Inquirer* reported the findings of the National Transportation Safety Board's in the accident which killed Governor Mel Carnahan and found that the pilot, Carnahan's son "Randy became disoriented and lost control of the Cessna 335, in part because the key instrument guiding him through darkness, rain, and fog malfunctioned." The news article reported that the "Investigators focused on the main instrument Randy Carnahan used to navigate, the primary attitude indicator. Also called the artificial horizon, it reports a plane's position in the air, telling whether it is banking and whether the nose is high or low. NTSB investigators concluded that the indicator 'was not displaying properly at the time of impact. A smaller secondary attitude indicator was working, they concluded. Because it was smaller and several feet away, Randy Carnahan would have had to turn his head frequently and rapidly for cross-checks with other instruments... Defendants have said that negligence by Randy Carnahan led to the crash." Determinations in such cases are best left to the experts, since, even at best, they have access to more information than sideline judges and have the responsibility to put their observations together and come up with their findings.

With the reliability of both instruments and electrical power supplies today, the likelihood of having to fly an aircraft using only the "partial panel" instruments and maneuvers just described is very rare. Today, it is likely that partial panel skills are not stressed so much. But if the pilot finds himself in a situation in which such use is necessary, he can control the aircraft very safely; and, given certain conditions and circumstances, bring the aircraft to a safe landing in bad weather. He would, of course, need to have some electrical power for his radio — mainly for radio reception to be talked down as in a GCA instrument approach to landing.

And finally, instrument flying is required by Naval Aviators in other instances when the weather is not bad. Some nights at sea are just plain black when there is no moon and when the stars are obscured by overcast skies and when haze or mist severely limits visibility. At some times in these conditions, there are not any surface vessels either to provide reference points. In such an environment, there is no horizon, so there is no <u>left</u> or <u>right</u> or <u>up</u> or <u>down</u>. For the pilot, it is just easier and simpler to fly "on instruments." And even in bright, sunny, clear weather with unlimited visibility, it is just easier for the pilot to fly "on instruments," yet keep an eye out for other aircraft. Few pilots can fly the aircraft as precisely by reference to the actual distant horizon, as they can by reference to the aircraft's artificial horizon and other instruments. Further, using the instruments in such manner makes it quite easy then for the pilot to engage the autopilot, or as some pilots say "Put it (the aircraft) on George," and, as they say "Let George do it."

Instrument Flying at Its Best: The "Pure Fun" of It!

Lieutenants Karl Laufer and Dan Preston were hardly aware of the moderately heavy rain falling outside their taxi as it sped towards the Naval Air Facility at Lambert Field in St. Louis, Missouri, this early April morning. They were most concerned that, in the heavy traffic they were in, they arrive at the field in sufficient time before their VIP passengers in time to get the aircraft inspected and ready to depart on time. Having been in the West the day before, they were now to carry out their assignment of picking up some Navy Medical VIP's and returning them to the Naval Auxiliary Air Station, Mustin Field, Philadelphia, their own base of operations. Their passengers were to be Commodore Kern, a reserve officer and a well-known physician from Temple University's School of Medicine, Captain Schmidt, the Commanding Officer of the Philadelphia Naval Hospital, and a commander, who was the Executive Officer at the Philadelphia Naval Hospital. The medical officers had been attending the annual conference of the American College of Physicians and Surgeons in St. Louis and were now to return to their responsible positions.

The heavy traffic concerned Laufer and Preston because they had wanted to get to the Naval Air Facility at Lambert Field early, not only to preflight their aircraft but also to have time to check the weather en route and at their destination and to file an instrument flight plan with the Federal Aviation Administration (FAA). The instrument flight plan would assure them a specific altitude and airway route, something absolutely necessary if they were to be flying in the black clouds which had been producing the moderately heavy rain they were encountering en route to the airport. Their passengers were to arrive at Lambert Field at about 1100 hours Central Standard Time.

Their aircraft was a utility airplane ideally suited for the mission involved. Known simply as the Beechcraft or SNB, which was also called the JRB depending on its configuration, it was manufactured by the Beech Aircraft Company, who made many of these small airplanes for both the Navy and the Air Force. They were marvelous little airplanes which could carry two pilots and four passengers plus baggage; and they were very useful and economical in serving a variety of needs. Tricky to get to know well enough so that the pilot could avoid bouncing on landing, it was not everyone's wish to fly it. But for those who had mastered it, often after going through the bounces in their first few hours of flying and making practice landings, it became a pleasant experience and a nice change from the much bigger, much heavier operational military types. An inconvenience, albeit minor, was that the fuel was contained in four fuel tanks, two twenty-five gallon tanks and two seventy-eight gallon tanks, all of which had to be individually selected. The two hundred six gallons gave the aircraft a flight duration of about four hours, possibly five hours, depending on power settings. In some later models, a nose fuel tank with a capacity of forty-seven gallons had been added in a space formerly used for baggage. (Placing baggage in the nose cone area was not without its problems, however, as one very interesting experience related in the *Saturday Evening Post* would reveal.)

Relieved that their taxi had delivered them to Lambert Field by 0945 hours, the pilots placed their luggage in the airplane, checked their aeronautical charts to insure that they had all the necessary airway information, and preflighted their aircraft, inspecting carefully to insure that they had full fuel tanks and to look for anything unusual — and in the process, getting themselves pretty well drenched in the continuing moderate to heavy rain. These things completed, Laufer and Preston filled out their flight plan with the names of the pilots and passengers and including, most importantly, their desired route or airways and a preferred altitude which was more or less dictated by the limitations of the airplane and its equipment. But before submitting the flight plan to the Naval Air Facility's operations officer for clearance, and considering that their clearance would be automatic, Laufer and Preston proceeded to the Aerology Department for a weather briefing. Looking at a large, recently compiled weather map, they could see a huge, wide green area in the shape of a broad peninsula from well north of St. Louis to well south, seemingly to near the Gulf of Mexico. The huge green area indicated precipitation, no doubt rain at the surface this time of year, but who knows what at higher altitudes. St. Louis was the center of the green area's east — west dimension, the green area extending about two hundred fifty miles to the east and a somewhat similar distance to the west. Neither the rain nor the extent of it was of particular concern to Laufer; but with a surface temperature of below fifty degrees Fahrenheit, a flight level of as low as five thousand feet could bring freezing temperatures, snow, and ice instead of rain. Knowing that the air temperature decreases from three to five degrees per thousand feet of altitude, the variation dependent on certain atmospheric conditions, Laufer knew that he would have to employ the deicing equipment in flight. The Beech had fairly good anti-icing equipment; and Laufer knew that, properly utilized, it would be more than adequate to combat most icing conditions and certainly for the conditions which their aerological information would indicate — provided, of course, that it would function as designed when needed.

The Commodore and his fellow doctors arrived a little late because persistent rain has a way of making everything take a little longer. Knowing that his passengers were ready to go and that they would just naturally expect to go, Laufer submitted his flight plan to the operations officer for approval, who in turn would forward his request for preferred airway routing and altitude to the Federal Aviation Administration — and also for the formality of requesting permission to depart from Lambert Field. In considering how to file his flight plan, Laufer had calculated that the miles en route to Philadelphia were about eight hundred twenty nautical miles (nine hundred twenty statute miles) and that the normal cruising speed of one hundred fifty miles per hour would mean that, with an expected tail wind, about five hours and thirty minutes of flight would be required. The flight time could be less if the tail wind was stronger than expected. With fuel being consumed at the rate of about forty-five gallons per hour, they would need to have about two hundred fifty-five gallons of fuel. These calculations did not include the fuel that would be consumed in start-up, taxiing, engine run-up during the takeoff check, and the fuel used while waiting for the instrument flight clearance from the FAA. And, more importantly, these calculations did not include the extra fuel that would be consumed with their high power settings while they were climbing to altitude and getting established on course. In addition to these considerations, there would have to be another: to have enough fuel to fly to an alternative airfield if Mustin Field in Philadelphia was below the minimum weather conditions for landing. Even if the aircraft had been equipped with a forty-seven gallon nose tank, they would still not have been able to make the entire trip non-stop.

Laufer had made these calculations earlier, deciding that a stop en route would be necessary and chose to stop at Columbus, Ohio, a field he was well-acquainted with and whose Naval Air Facility would be able to refuel them quickly if time should be a factor. All that was necessary now was for Laufer to wait momentarily for the operations officer to sign and return a copy of his approved flight plan while he kept the original to forward the routing request to the FAA. Very much to Laufer's surprise and somewhat to his dismay, the operations officer wanted to see Laufer's instrument rating card to verify Laufer's instrument rating and to see that it was current. Laufer had never experienced such a request from an operations officer. Then, after seeing Laufer's instrument rating card, surprisingly — if not strangely — the operations officer told Laufer that his aircraft was overloaded. Laufer was puzzled because he knew that, although the aircraft was approaching its maximum gross weight, it was not overloaded — particularly since there were only five people aboard instead of the six it could carry. The operations officer claimed that the luggage more than made up for a sixth person. Laufer had calculated roughly that the combined luggage weighed no more than 300 pounds, that adding 1,236 pounds for the two hundred six gallons of fuel, and that figuring each passenger at a generous 180 pounds added up to 2436 pounds. Adding that to the aircraft's basic weight of 6,320 pounds totaled 8,756 pounds. Since the maximum allowable gross weight was 8,730, the operations officer was technically correct, if not far-thinking. The aircraft was overweight by about 26 pounds or about four gallons of gasoline, an amount that would be reduced very rapidly after start-up. Perhaps the operations officer was expressing concern for the weather conditions; and perhaps there had been some incident at the Naval Air Facility at Lambert Field recently which prompted his concern and a closer surveillance of instrument flight clearances in bad weather conditions. Perhaps, in view of the VIP passengers, he wanted to cover his own ass. Or perhaps his actions were the result of his own feelings of inadequacy. Whatever, it would really be difficult to know, and Laufer did not want to think about it now. So Laufer and Preston would never know, and they certainly did not need any additional concerns. But unfortunately, the operations officer's remarks had an effect on Dan Preston, who became worried and thought that the flight should be postponed and scheduled for the next day when the weather was forecast to be much better. Now, Laufer had to deal with two things: both the operations officer and his copilot Dan Preston. Laufer countered Preston, reminding Preston that he, Laufer, was well-qualified for instrument flight, not superficially qualified, that the aircraft was instrument-certified and had good deicing equipment, that there was no justifiable reason for not going, and that they had a compelling reason, under the circumstances, "to deliver the goods." To Laufer, it was also a matter of pride, not of false pride, but something that said, "If you're qualified, you're supposed to perform." It was as simple as that. To Laufer, after all, military flying was not like taking your grandmother up for a flight in a small airplane on a sunny Sunday afternoon. If it was to be effective, it had to be around-the-clock and in all weather. And if that was the case, you could not beg off with half-baked excuses. It was not that Laufer was reckless; on the contrary, it was a matter of knowing yourself and having respect for the job.

After reassuring a not-too-easily reassured Preston, Laufer then went to the operations officer. Considerably annoyed for what he considered unnecessary interference, Laufer knew that there was an easier way than fighting with him. He told the operations officer in a somewhat conciliatory manner that, given the existing weather conditions, there would surely be some delay between the time they started the engines and the time they would taxi out and warm-up, and given the time they would wait to receive their routing and clearance from the FAA and would be cleared for takeoff, the delay would consume enough of their fuel to insure that they were below the aircraft's maximum allowable gross weight of 8730 pounds. He also reminded the operations officer that they had no nose fuel tank and that, had they had a nose tank of fuel, they would certainly be overloaded and would have to siphon fuel from it. Now aware of Laufer's knowledge and confidence, the operations officer could hardly do other that sign the flight plan — or was he now sure that he had satisfactorily covered himself?

After taxiing to the edge of the takeoff runway and going through their engine run-up and preflight check-off list, Laufer and Preston sat back to await their FAA instrument clearance, not really knowing how much of a wait to expect. But surprisingly after about ten minutes, they received their FAA clearance, which pretty much assigned them their preferred airways routing, their assigned altitude of seven thousand feet and some radio reporting instructions for just after takeoff and during their climb to seven thousand feet. Laufer was aware that seven thousand feet might put them into some "interesting" icing conditions; he, nevertheless, accepted the clearance, and, as required, read back verbatim the FAA clearance so that the FAA would know that they had received the routing precisely. They were then cleared for immediate takeoff and taxied onto the active runway and advanced their throttles. At the recommended throttle settings, they released their brakes and advanced the throttles to full power as they moved down the runway gathering speed in the continuing moderate to heavy rain. The aircraft handled nicely, confirming Laufer's opinion; and if there was any sluggishness due to their near maximum gross weight, it was hardly noticeable. The aircraft seemed airborne in about the usual runway space; and in seconds they were in the clouds at four hundred feet, climbing "on instruments" to seven thousand feet, where they leveled off in the dark clouds where they would remain for two hours, and on their assigned airway in a very precise manner. Had they been in a jet or in a high-performance, pressurized airplane, they could probably have climbed quickly through the clouds and into the bright sunshine above. But they were not, so they had to be ready for what "fun" might lie ahead — fun which would come rather quickly.

During the takeoff and climb to cruising altitude, Laufer was generally aware of how the rain started to be mixed with snow, until it seemed to change all to snow, a soft wet slushy snow, which ordinarily should not present any serious problems. Although Laufer had done his share of instrument flying through some pretty severe weather, he did not recall that he had had much experience in these conditions, conditions which could change for the worse as they proceeded en route. He did, however, remember well those things the Navy had taught him about icing; and he had confidence in that information, so he felt prepared to deal with whatever would come in his present situation. His main thought was to try to anticipate changes in conditions and to try to be prepared for them: if he flew into colder conditions, would the slushy snow, which now only built up to a certain point on the wings before being blown off by the wind stream, start to freeze hard on the wings, tail surfaces, and propellers? A buildup of hard snow or ice would change the shape of the wings and affect the airflow over the wings, reducing lift and requiring more power to maintain altitude. If they should fly into colder air, though unlikely, and then encounter rain which then froze, there would be a clear ice build up over the upper surface of the aircraft, adding weight dramatically and stubbornly resisting any of the anti-icing equipment on the aircraft. Laufer had encountered freezing rain or "clear ice" before, knowing that it could build up quickly and add weight to the airframe. He knew that landing quickly would limit the exposure to the ice buildup, or, if that was not possible, then the only solution would be to change the en route altitude to a higher level where the air would be warmer — upward to a warmer air level occasioned by a temperature inversion. After all, precipitation as rain comes from a warm level and becomes clear ice (freezing rain) only when it strikes objects which are at or below a freezing temperature. Perhaps Laufer thought of these weather circumstances because he remembered an incident when he was flying into Columbus a year earlier when what looked like rain striking his windshield was actually freezing rain. Although the rain did not completely freeze on the windshield because of cabin heat, it did freeze solid as it struck the cold, below freezing metal parts of the aircraft. Because of the nature of the clear ice, Laufer had not noticed how the ice was building up on much of the aircraft; and by the time he landed at Columbus five minutes later, the freezing rain had built up to a thickness of one quarter of an inch — all over the upper half of the airplane being pelted by the rain. At that time, Laufer was not able to calculate the weight of the ice on the aircraft; but he felt that it must have been considerable and wondered how much more weight the aircraft could carry. The worst part of that experience, Laufer remembered, was that he could not see how rapidly the buildup had occurred until after he had landed. Of one thing he was sure: that he was glad he did not have to fly a greater distance in those conditions at that time.

But today, things were much different as Laufer could easily see from the slushy buildup on the leading edge of both the left and right wings, slush that would only build up to a certain depth before sliding off in the slipstream constantly flowing over the wings. If conditions did not deteriorate, there would probably be no need to employ the wing deicers, expandable leading edge boots designed to break into pieces any buildup of either snow or ice, which would then be blown off as the slushy snow was now being forced off. Even the employment of the expandable wing deicer boots had to be effected with caution. Employing the boots constantly could create a hollow space in front of which the ice or snow could accumulate. The expansion and contraction of the boots within this hollow would be as bad as not using the boots at all, the accumulating snow or ice eventually distorting the smooth flow of air across the wing and diminishing the lift. Established as desired on course and at the assigned altitude, Laufer double-checked the Pitot tube heat to keep the tube from filling up with the slushy snow and ice. The Pitot tube, with its "L" shaped tube extending down from the underside of one wing and its open-ended front tube which faced into the direction of flight, permitted the ram air to come into the tube to indicate airspeed; but its open-endedness likewise permitted rain, ice, and snow to enter as well. The Pitot tube also had a static vent or hole which allowed air pressure to be measured and thereby provide altitude information. Without Pitot heat and in conditions like those they were experiencing, the Pitot tube would freeze up, rendering the crucial air speed indicator and altimeter useless. Anyone who had experienced a frozen Pitot tube and seen its effects on his instruments would not be likely to forget it. Fortunately, however, even if a Pitot tube freezes up, it is usually unfrozen with the application of Pitot heat. But it is not too difficult to imagine a bad situation in which a pilot is making an instrument letdown in bad weather and the tube freezes up while the pilot is in the clouds and getting close to the ground.

Laufer knew that he would have to monitor all of his instruments closely and, of course, continuously in the present, and perhaps changing weather conditions. Preston too was to keep a close watch. But Laufer did want to check also the cabin area to see how comfortable the VIP passengers were. He turned and looked back to how they were doing and what, if anything, they might be engaged in like reading or composing notes from their meeting. What he saw was one of those serious yet comical scenes that in life frequently occur together. The green upholstery material which lined the interior walls or bulkheads of the cabin had huge wet splotches making the green darker where the rain and slush had been forced into the cabin through the riveted seams in the fuselage. Such was the volume and force of the rain and slush pelting the aircraft. As a result, the dignified commodore, the captain, and the commander had to lean as far away from the bulkheads as possible to keep from getting their formal dress-blue uniforms soaked. Dignity had vanished, now to be replaced by practicality.

Unable to do much to change the "damp" conditions aft in the cabin except for increasing the cabin heat, Laufer turned his attention once again to the cockpit, only to be greeted by a loud cracking sound which came from the starboard side of the aircraft low and to the front. Startled for a moment, he soon realized the sound came from ice which had built up on the spinning propeller blades on the starboard engine and was being hurled against the lower front fuselage by the centrifugal force of the turning blades. The sound soon came from the port side; and it was to occur many times during the flight. The aircraft was equipped with propeller anti-icing fluid, which they used; but it was used a bit sparingly in view of the existing, mainly slushy conditions. Relatively assured now that the wing and propeller ice presented no serious problems in the existing conditions, and aware that they could not control the amount of rain/slush coming into the aft cabin, Laufer and Preston started to review again the instrument letdown charts for Columbus, when suddenly the starboard engine began to cut out badly. Though again momentarily surprised, Laufer knew precisely what the problem was and instantly boosted the amount of heat to the right engine carburetor, causing the engine to respond immediately and resume its power and purr smoothly. Both Laufer and Preston were briefly startled because they had been using carburetor heat from just after throttling their engines back after takeoff, the need for more heat reflecting colder temperatures aloft. The design and construction of carburetors is such that carburetor ice seems a natural consequence, so all pilots are aware that they must use carburetor heat and, in addition, take special precautions in conditions like those which Laufer and Preston had encountered. After the cutting out of the starboard engine due to carburetor ice and their applying more heat to that engine, Laufer turned his attention to the port engine, only to find that, before he could actually apply more heat to it, it too sputtered a bit before it was given the same treatment and restored to good health.

For such a small aircraft, there were quite a few anti-icing and heating systems. Besides cockpit and cabin heat, which pilots know really are essential if they have ever taken off on a bitter cold day and have nearly been numbed and incapacitated when their cockpit or cabin heaters did not work, there were other anti-icing systems crucial to flight in conditions like those described: wing and tail deicer expansion boots; propeller anti-icing fluid; carburetor heaters; Pitot heaters; and windshield heat to keep ice from forming and blocking forward vision. It was, of course, the presence of all of this equipment that made this aircraft an all-weather certified aircraft, one that would permit Laufer to undertake this flight with a degree of confidence.

With the engines running smoothly and the propellers and wings staying clear of ice, once again they turned to the task of reviewing and rehearsing the letdown procedures for Columbus Airport. But with still twenty minutes to go before reaching Columbus, they were surprised to find themselves out of the clouds at their flight level and in the clear. The higher clouds remained; and the precipitation which had consisted moments earlier largely of slushy snow, now seemed to be all rain, still a moderately heavy rain. Although the aircraft's being in the clear was of no help to the passengers, the moisture continuing to seep into their cabin, it made things a little more simple for the pilots. Under the circumstances and in the interests of saving about ten minutes time, the time it would take to go through the full, formal instrument letdown procedure, they decided to cancel their instrument flight plan and head directly for the airport and landing. They called the FAA to cancel their flight plan and then contacted the tower at Columbus for landing instructions. The flight, though interesting so far, had presented them with no problems or situations they did not expect and for which they had not been well-trained. They were not, however, prepared for what they were now about to encounter.

After getting landing instructions, letting down to the prescribed altitude and landing pattern, and making their approach turn towards the runway in use, they noticed that, as they reduced their approach speed, the rain was not being blown off their sloping windshield very well. Pretty soon it was piling up so that, even though light was coming through the windshield, there was actually no forward visibility. Just a hundred or so feet above the ground and approaching the runway fast, they faced a situation for which no training had been or could have been provided: how to land safely when the windshield is completely obscured by the accumulation of heavy rain piled up on the windshield whose slope just counterbalanced the force of the slipstream during the slow speed on the final approach. Under such circumstances, it is just impossible to see the horizon or the runway ahead on which the landing must be made. Laufer knew of no other military aircraft whose characteristics were such that the rain could be so contained. Many aircraft had windshield wipers; and wipers were indeed installed on later models of the SNB. And even if wipers were available, there is no indication that they would have been of use — as anyone who has driven in a torrential downpour, even using windshield wipers on the "fast" cycle will attest. But in the SNB it was much worse; and they could not worry about the absence of wipers now, realizing that they had to get on with it. No doubt, other pilots with Laufer's flight time in the SNB had experienced similar situations — and Laufer had too, on a flight going into Cleveland Municipal Airport on one occasion. Although Laufer knew what had to be done if they were to land, was it really fortunate that he had faced a nearly identical situation in landing satisfactorily at Cleveland? He knew all too well that having been successful once before was no guarantee that he would be successful this time.

At any rate, Laufer was determined to make an attempt and so informed Preston, who did not seemed cheered by the news. Laufer knew full-well that he would have to be very careful, so that, if necessary, he could take a wave-off and either wait awhile and hope for the rain to lessen or proceed to an alternate airport. Where the gently sloping windshield curves to the side and blends in with the side windows, there is a small triangular-shaped window of about eight inches at it widest which runs vertically for about ten inches and which can be opened inwardly. Looking out through this now-opened window nearly at right angle to the direction of flight down the runway, Laufer could judge his height above the ground and could keep his wings level, but he could only see as far forward as the propeller on the port engine. To him, it seemed his forward view extended only about twenty-five feet, a distance totally inadequate for a safe landing.

As at Cleveland, when Laufer was sure he had reached the threshold of the runway, he focused his attention outside the small open window on his port side, keeping his left hand on the control yoke which controlled the ailerons and the tail elevator controls and his right hand on the throttles, also occasionally checking his airspeed indicator. Noting that there was hardly any drift either to right or left from a crosswind, Laufer made sure his wings were level and planned to ease the plane down close to the runway surface. Looking out the same small window on the port side, he adjusted his throttles slightly back and kept holding the aircraft just above the ground, hoping to let his speed decrease and the aircraft ease down to touch the runway. Just as they did in Cleveland, the two main wheels touched the ground in a surprisingly gentle manner. At about the same time, and with the aircraft's speed dropping quickly, the windshield cleared, enabling Laufer to look down the runway for the first time and to ensure that the aircraft maintained a straight course down the runway. Thankful for his experience in Cleveland, Laufer felt that that made this approach and landing less of an unknown quantity.

They quickly taxied to the Naval Air Facility for parking and refueling. The flight to Columbus, the first leg of the trip, was over. Their getting airborne later than Laufer had expected because of the VIP's somewhat late arrival at Lambert Field in St. Louis, the time delay occasioned by their getting the weather briefing, the delay caused by Lambert Field's operations officer, and the wait at the edge of the runway for their FAA instrument flight plan meant that their elapsed time was more than Laufer had anticipated. In addition, since they had crossed into Eastern Standard Time, it was an hour later. So by the time everyone got out of the plane in Columbus, it was nearly 1500 hours. But Laufer was not concerned because he himself was not in any hurry. His main concern was for his passengers and that he get them to Philadelphia safely and in a reasonably convenient time frame. Preston told the VIP's where they could get some light lunch, while the aircraft was refueled; and then he and Laufer too went for a sandwich and a cup of coffee. Very much in awe of their passengers' rank, Laufer and Preston did not join them for lunch; but they did advise them respectfully to take however much time they wished and that they would be ready to depart for Philadelphia whenever they returned to the aircraft.

Uncertain about how much time the commodore and the captain would want to take for lunch, Laufer and Preston thought that they had better get on with filing another instrument flight plan, checking the weather both en route and at the destination, and once again giving the aircraft a good inspection to make sure that there was no residual ice anywhere on the aircraft which could freeze at altitude. They could do nothing about drying the soaked tapestry on the aircraft's bulkheads to make their passengers more comfortable, so they did not give the idea much thought. With everything checked and ready, Laufer and Preston had only to wait for their passengers to appear before they submitted their flight plan to the operations officer. The VIP's had a relaxing lunch, discussing no doubt something about matters medical, Laufer thought, and forgetting that an hour would be lost in going from Central Standard Time to Eastern Standard Time. (It would be a year later before Summer Daylight Savings Time would start on April 1.) Nevertheless, they took their time, and they appeared at the aircraft after an hour at lunch. Although the aircraft weight, etcetera, was exactly as at Lambert Field, the Columbus Naval Air Facility's operations officer had no doubts or questions concerning whether the aircraft was overloaded, correctly signing the flight plan and wishing the pilots a good trip. The Columbus operations officer's prompt clearance of their flight reinforced Laufer's concern about the Lambert Field Naval Air Facility's operations officer's hesitation. Without delay then, they taxied out to the runway, ran up the engines, received their instrument flight plan, and were airborne again at 1630 hours EST. The weather forecast was for continuing moderate rain, diminishing farther to the east. Although they probably could have made the flight under visual flight rules, Laufer and Preston wanted to continue under instrument flight rules to insure a place on an airway and an assigned altitude, in other words, under the positive control of the FAA and the continuous surveillance of its radar. It would insure that, if by chance they encountered "below minimums" weather again, they could proceed without the delay of their having to request an instrument flight clearance while en route, a delay that might mean circling or holding in better weather until FAA could approve their request, and perhaps assigning a more circuitous route and another altitude. Hence an instrument flight plan from Columbus to Philadelphia was preferred.

Shortly after takeoff, the rain had all but stopped, and likewise the threat of icing except for carburetor ice, an ever-present consideration. By about 1800 hours, the last rays of the sun had disappeared, and, surprisingly, stars started to appear. And by 1830 hours, LT Laufer touched down smoothly on a very dry runway at the Naval Auxiliary Air Station, Mustin Field, at the Philadelphia Naval Base, having been able to see through a clear windshield and thinking to himself how nice and easy it is to fly in relatively normal conditions. Even if the second half of the trip was relatively boring by comparison, Laufer had to admit that a break from the experiences of the first half of the trip was not entirely unwelcome.

Many times, senior officers not in a junior officer's direct line of command have very little to say to junior officers; or sometimes they give superficial recognition, purposely keeping a distance dictated by the difference between their rank and that of the pilots. But the Commodore and his fellow officers seemed particularly friendly and appreciative, expressing sincere thanks for an experience that extended from nearly monsoonal wetness, snow, rain and ice while in flight to a smooth, gentle landing on a dry runway on a beautiful starlit night. There was little likelihood that they knew anything about the blind landing in Columbus; but, nevertheless, they seemed not to mind "hanging around" and talking about the flight, asking questions while they lingered in their still-damp uniforms. No doubt, this was a new experience for the VIP's, one they seemed pleased to have had, and they appeared to want to take a few minutes to put all the pieces in place. No doubt, the small size of the SNB made their immersion in the "interesting" weather a much more intimate experience. To Laufer and Preston, such interest and treatment from seniors officers made this not-uncommon trip something special. They too would remember the VIP's and this trip for a long, long time, longer for Laufer since LT Preston was killed a few years later while making an approach turn to land on an aircraft carrier. But as for this instrument flight, it was "Instrument Flying At Its Best."

Not a sophisticated aircraft, the SNB was a good airplane to fly, mostly as a pleasant change from bigger, heavier, faster aircraft propelled either by jet or reciprocating engines. It really was a good utility aircraft capable of accomplishing many tasks which would be inappropriate for other aircraft due to their high cost of operation. It also gave pilots a true sense of flying. It didn't take you above the clouds and weather; it put you right in the weather. Its flight controls were cable-connected and not hydraulically boosted, providing a realistic sense of feel. It was this sense of feel that enabled LT Laufer to attempt the landing with the rain-obscured windshield, even if such an attempt might be considered unwise by many others. For the reader who may have thought that a Ground Controlled Approach (GCA) would have lessened the hazard of Laufer's landing, it should be remembered that the GCA controllers advise the pilot to take over visually at the end of the runway, while still airborne, to make the landing.

The problems and hazards of icing on aircraft constitute a huge category of incidents. The reader has seen here just one example of icing and not severe icing at that. But perhaps it has served to explain the general nature of icing. For a more detailed "unbelievable" description of an icing incident, the reader might want to read about an example described in Ernest K. Gann's book *Fate Is the Hunter*.

NAS South Weymouth, MA, to NAS Willow Grove, PA

There was no rain, and the visibility at ground level was good that day at the South Weymouth Naval Air Station in Eastern Massachusetts. There was a low overcast of clouds, however, which extended from 400 feet solidly up to fifteen or sixteen thousand feet. The squadron of S2F Trackers, carrier antisubmarine aircraft, was scheduled to return to the Willow Grove Naval Air Station in Pennsylvania near Philadelphia, the next day — a Friday. The pilots in this reserve squadron were nearing the end of two weeks of active duty for training, during which each pilot had flown about sixty hours. There was no reason the pilots could not return to Willow Grove late Thursday afternoon, and a few wanted to because of work that was waiting for them back there. It may have been that a bit of inertia had set in; but most pilots wanted to wait until Friday, thinking it would hardly be worth the effort of filing an instrument flight plan with the Federal Aviation Administration close to the end of the business day and then waiting in the airplane with the engines turning while waiting for Federal Aviation Administration approval and the assignment of a specific route and altitude to be flown to get to Willow Grove. An important factor in their thinking was that the route from South Weymouth to Willow Grove would be in the heavily congested air corridor between Boston and Washington and which included traffic through New York, Philadelphia, and Baltimore as well.

One pilot, Lieutenant Commander Dutch Heilman wanted to try getting out because he could use the full day at Willow Grove on Friday. Also, needing an instrument flight proficiency check, he felt that their flight would be in the clouds the entire way and would provide realistic conditions for the flight check. Besides, there was no assurance that Friday's weather would be any better. The only thing better would be that, instead of getting airborne at about four o'clock on Thursday, they could get an early start after a good night's sleep. But Heilman was determined, so he convinced the instrument flight check pilot, Commander Hal Wingate, that they should go immediately. Whether the enthusiasm Heilman exhibited had any effect on others, he did not know, because no others decided likewise. And it was just as well, because the events which followed became a lesson to remember, not to mention a dislike of the FAA because of what Heilman thought might be their arbitrary and capricious actions that Heilman's attempt to get a timely and acceptable route and altitude would foster.

Alerting his four man crew, Heilman filed his instrument flight plan and they taxied out to the takeoff end of the runway in use to perform their takeoff check list and wait for the FAA clearance they felt was imminent. After about twenty-five minutes they received their route and altitude assignment from the FAA people, being somewhat shocked to find that the assigned altitude was to be thirteen thousand feet when they had asked for seven thousand feet. Two things became clear very quickly: The traffic in the corridor between Boston and Washington must be extremely heavy; and their assigned altitude of thirteen thousand feet would have to be refused. Basically a low altitude antisubmarine aircraft to be utilized mainly in warfare against submarines, the S2F was not equipped with oxygen, nor was the cabin pressurized. While the aircraft could easily operate at thirteen thousand feet and beyond, flights above ten thousand feet in daytime were prohibited if no oxygen was available or if the cabin was not pressurized; at night, the restrictions without oxygen were more limiting. The pilots could probably have functioned at thirteen thousand, but they would be ignoring the safety rules which the Navy had so widely promulgated and which they had learned so well. Regretfully, they advised the FAA that they would have to wait for a lower assigned altitude. And wait they did! After leaning out their fuel-air mixture to avoid carbon buildup in the engines from the full-rich mixture used for takeoff and after an hour of waiting, they called the FAA to see how their request was coming; FAA responded that they were "working on it." Another hour passed, the engines still cranking over, and the pilots were thinking that another clearance with a lower altitude would come at any moment. Another call to the FAA prompted an identical response. Heilman revved up the engines to help clear out any buildup of carbon and leaned out the air-gasoline mixture again just to insure no carbon would build up in the cylinders. Another half hour and another call to FAA — and a similar response with an additional comment that they could not give them a predicted clearance time for a flight at ten thousand feet or below. No doubt, air traffic was heavy and not easing at this time of day; and it was likely that the busy and stressed out controllers had bigger concerns. But still a bit hopeful, Heilman and Wingate waited — until after three hours and fifteen minutes of idling their engines, they started to think that the FAA people were either deliberate, punishing them for refusing the original assignment of thirteen thousand feet, or that in their frantic condition they had simply forgotten them. And Heilman and Wingate wondered if they had been unwise in refusing thirteen thousand feet, because they felt that, once airborne and at thirteen thousand, they could then request a lower en route altitude. After all, they had flown many times out of El Paso and over the mountains toward Tuscon where an altitude above twelve thousand feet was necessary. They felt a lower altitude would be much easier to obtain once airborne and on their way. Dejected now that they had been delayed so long, they felt a certain resentment toward the FAA and resigned themselves to getting an early start Friday morning. They also felt they had wasted time and were no better off than those pilots who sat back and relaxed.

Friday morning the weather was the same as far as cloud coverage and visibility were concerned, but, in addition, a moderately heavy rain was falling. Undaunted by Thursday afternoon's events, Heilman, Wingate, and crew were the first to get their instrument flight plan filed with FAA and taxi to the end of the duty runway to go over their takeoff check list and wait for their assigned route and altitude. Thoughts of the delays of yesterday afternoon raised doubts about what today's altitude assignment would be. They needn't have worried. Whether there was less traffic in the Boston to Washington corridor at this hour of the day — as seemed likely — or whether there was a Navy-friendly FAA operator they would never know; but surprisingly their requested route and altitude were granted at the moment they had completed their check-off list, and they were given immediate takeoff clearance. After they climbed to their assigned altitude and established themselves on the assigned airway, Heilman thought about how lucky they were to get airborne so quickly. Though a pleasant surprise, Heilman was hoping that the prompt approval of their flight plan would be the only surprise that morning, realizing fully that, at any time, other surprises could follow. En route, they were constantly in the clouds, experiencing some turbulence and hard heavy rain. Icing at this time of year would not be a problem. Everything proceeded exceptionally well, and the only unknown of any concern was if the weather at Willow Grove Naval Air Station would remain reasonably close to the 400 feet of altitude and three miles visibility which had been forecast earlier. They also thought about what kind of instrument approach Willow Grove would assign for them to get beneath the clouds and down to the runway. Most likely, they thought, would be a radar approach or GCA.

As they proceeded along the route, they could hear radio contacts of other squadron aircraft behind them, who were for awhile at least within the same FAA Control areas, as the pilots behind called FAA as they reached required reporting points. Wingate spoke on the intercom saying, "It looks as if everyone is getting out of South Weymouth." It was reassuring to know that all of the pilots and crews were getting out of South Weymouth in good order. Apparently, air traffic was much lighter at this time of day, and perhaps there was some record of yesterday's delays. An excellent active duty period so far, it looked as if there were no real problems and that they could wrap this training period up in the manner everyone wanted. And there was special merit in their being able to have such a good instrument flight experience in the clouds and heavy rain as one of their end-of-cruise flights.

As they neared Willow Grove, the turbulence seemed to diminish considerably, but the rain seemed heavier from the sounds it made against the fuselage. Then as expected, FAA called them advising them to contact Philadelphia Approach Control on 291.6 megahertz, who in turn advised them to contact Willow Grove Ground Controlled Approach (GCA) on 288.6 megahertz. Calling GCA, they were advised that GCA had them on its radar and that Willow Grove weather was overcast with clouds at 300 feet and that visibility was three miles. The combination of ceiling and visibility seemed quite satisfactory for a GCA radar approach; and Heilman, following radar instructions, descended to the assigned altitude and heading. He then maintained level flight, as directed, until he reached a spot called "the glide slope intersection." Prior to this point, the pilot had to repeat each instruction given by the GCA controller. But at this point, final controller takes over in a one-way conversation to bring the pilot down an electronically determined slope or path which leads the pilot down to a spot below the clouds where the pilot can take over visually to make a safe landing. But to stay on this slope, the pilot must follow precisely the voice instructions of the GCA controller who is in constant touch with the pilot, calling out changes in heading to compensate for the wind and instructing the pilot to make changes in his rate of descent if he is above or below the glide slope. The controller also keeps talking even if there are no changes necessary to the heading or rate of descent; in such situations, the controller will repeat the message "You're on the glide slope. You're on course." Or, "You're on Course. You're on the Glide Slope." And provided the pilot precisely follows the voice instructions of the GCA operators, he will be guided down below the clouds and in view of the end of the runway in use.

Having made hundreds of GCA approaches, this one was just one more; and, indeed, everything went very smoothly and routinely — except for one thing: As they broke below the cloud layer, the GCA controller said, "You are over the end of the runway. Take over visually." But Heilman could see no runway. Sure enough, the air field was in front of him, but all that he could see was the green grass of the airfield, made greener by all the rain. The GCA controller's approach line had Heilman and Wingate lined up to the left of the runway. From the right side, Wingate could see the runway and asked, Can you see the runway?" Heilman replied, No!" and Wingate said it's off to the right. Shall I take it in?" Without answering and still descending slowly, Heilman made a quick flip-turn to the right, saw the runway, made a correcting left flip-turn to line up with the runway center line, and in a second or two was in a position where he felt his wheels should be touching the runway. So low to the runway seemed the aircraft that Heilman wondered for a split second whether he had lowered his wheels. It took but an instant longer for him to realize that their landing had been so smooth that they could not feel the actual touchdown. The runway was not as smooth along its entire length, as the roll-out down the runway would soon demonstrate; but at the touchdown point, the very gentle landing, gave them pause for a moment or two. The smooth landing was a bit inconsistent with the required, last-second flip-turn corrections; but it, nevertheless, provided a nice finish to a good flight. Now, even the resentment

Instrument Flight

Heilman felt towards the FAA people the day before was lessened considerably. He really did know what the FAA people had to deal with in such a heavily traveled corridor as they were flying in and felt that maybe his irritation with them was unjustified. The intense traffic and the tensions generated by it were often evident in the stressful tone of the air controllers' voices and their hurried manner of speaking.

Ground Controlled Approach (GCA) has been a marvelous means of guiding pilots down to landings in very bad weather conditions. However, like most things where humans are involved, it is not perfect. If not perfect, it should not be inferred that an occasional slight deviation makes it unreliable. Most pilots, perhaps all pilots, think of it as a wonderful system. Just knowing that someone has you on his radar scope and is talking to you to help you down at your destination through some very bad weather provides a very comforting feeling. In this case, being brought down to the left of the runway presented no serious problems. In much reduced visibility, it could have presented difficulties requiring the pilots to execute a "missed approach procedure" and another approach. Neither Wingate nor Heilman gave it much thought, preferring to register it in the back of their minds for future approaches.

How many times GCA equipment is not properly aligned so that aircraft are not brought in on the center line of a runway is not known. Most likely, such misalignment is quite rare. This author has only experienced it twice; and if other pilots have ever experienced it, no one seems to have talked about it — probably an indication that such misalignment was really of no consequence. Knowing its reliability and the comfort it affords to pilots being "talked down" to a safe landing on the runway, it is very unlikely that it has ever been seriously misaligned. More serious is the problem caused by the need to shift the GCA from one runway to another, because the runway has been closed due to an accident or some other unanticipated problem. Such was the case described in Part I, when a closed runway resulted in the GCA's not being able to provide precision approaches right down to the runway. Still, the GCA operators were able to provide assistance in what is called a planned position approach to bring the pilot over the airfield and in advising him both when he is over the field and the level or height of the cloud cover above the terrain. Even these less precise approaches provide very valuable assistance, and few are the number of pilots who would complain about having to make such an approach.

Reflecting on the events of both Friday and today, Heilman felt a bit guilty in having thought that the FAA people might have been vengeful in making his crew wait so long and eventually having to postpone their departure. He had flown through some control areas where the controllers seemed so harried that they spoke so quickly that it seemed they just wanted to get rid of you — no doubt because they had so much other air traffic to handle. He knew, of course, that that did not mean that they were not interested in you. Rather it meant they wanted to dispose of seemingly less critical traffic to be ready for heavier, perhaps potentially conflicting traffic. Heilman knew also that a big part of the problem had been the aircraft he was flying. Primarily designed for operating from aircraft carriers and for hunting and finding submarines at low altitudes, it required no oxygen nor cabin pressurization. Its features served it well for this purpose; and while there was no reason to preclude its use of the airways, in some areas where higher assigned routes were helpful in handling dense traffic, its limitations became apparent.

The Distractions of Smog and Catalina Island Near Los Angeles

It is relatively rare that pilots have the same kind of surprising or challenging experience twice, but Lieutenant Commander Heilman had a very similar experience a few years later. This time he was flying a four-engined SP2E (formerly P2V-5) Neptune long-range patrol aircraft and had a different copilot and crew. His destination was the Los Alamitos Naval Air Station in the southern part of Los Angeles and his copilot was Lieutenant Commander Swede Hanson, who was his regular squadron copilot. They had a good professional and personal rapport and had devised excellent and disciplined cockpit communication procedures.

Having made a refueling stop at Nellis Air Force Base in Nevada, they were approached by the operations officer about dropping off some cargo at Luke Air Force Base near Phoenix. Since time was not a crucial factor, they agreed. The long length of the SP2E seemed ideal for the size of the cargo to be carried. Besides, neither Heilman nor Hanson had ever been at Luke AFB and felt that the stop might be interesting and informative. It did prove interesting in that at Luke there were many F-104 Starfighters being used to train German Air Force pilots. But Heilman and Hanson knew that they had to keep moving, so they made the necessary preparations for clearance and takeoff. Before long, they had taken off, expecting to land at NAS Los Alamitos in mid-afternoon. They had filed an instrument flight plan with the FAA despite what appeared to be beautiful weather en route, because they were aware that the Los Angeles area was frequently affected by smog that reduced visibility below that required for visual flight. Also, whenever possible, it was squadron policy to practice instrument flying including the filing of instrument flight plans with FAA. Besides, if you were flying an instrument flight plan, you were under positive control of the FAA's extensive radar during your entire route. Nearing Los Angeles, they were advised by Los Angeles Approach Control that, due to the now dense smog and air traffic, they would have to be routed by vectors or headings to space out the traffic, and would eventually be vectored to a position where Los Alamitos' GCA could work with them. Like practically everyone else, Heilman and Hansen had heard about and were very much aware of Los Angeles' smog, which was caused by particles of industrial and automobile pollution trapped by westerly winds and the mountains partly ringing the city. Often, a temperature inversion with its higher temperatures aloft kept the normally higher temperatures at the lower levels — and the smog from rising upwards to higher elevations where it could escape and be dispersed. But Heilman and Hansen had never flown in the Los Angeles area, being stationed mainly on the East Coast and on Atlantic Fleet aircraft carriers. Though he had once flown into the San Francisco area, and often into the El Paso, Tuscon, and Las Vegas areas, Heilman had never had a requirement to fly to Los Angeles till this flight. This was a new experience for him; and when he heard that the smog was thick this day, he looked forward to experiencing it and all that entailed.

Heilman was curious and a little excited to be flying into such a new environment. Somewhat like fog, smog was nevertheless still different. In those moments when he could briefly take his eyes off the flight instruments and steal a look outside the cockpit, he found that he could see straight down to the ground from his four thousand foot altitude, while seemingly there was absolutely no forward or even slanted visibility in any direction. They were frequently advised of other air traffic at certain positions relative to their own position, and particularly when they were given a new heading, but they could never see any of it. Closely adhering to the vectors assigned by Los Angeles Center and precisely changing to new vectors when assigned, Heilman continued to steal looks outside the cockpit, most often and surprisingly seeing only ocean straight down; during one of these looks, he saw a big island which he figured had to be Catalina Island, an island about twenty-seven miles southwest of Los Alamitos NAS. This was the first time since they had contacted Approach Control and had been receiving vectors that they knew exactly where they were. Fascinated by seeing an area and an island he had heard so much about and which had such a fascinating and romantic sound, he seemed to have a juvenile delight in passing over it — even if he only had a straight down view. He was at the same time very careful to adhere to the vector instructions. So interested was he that it seemed only a minute or two before he was advised to contact Naval Air Station Los Alamitos' Ground Controlled Approach (GCA).

Hansen contacted Los Alamitos' GCA, and GCA replied promptly that they had the aircraft on their radar scope and would vector them toward the glide slope and the proper altitude at which to intercept the glide slope and then to begin their descent on the assigned radar heading. In a few moments, they were at the intersection of the glide slope and were advised by GCA that they were "on the glide slope and on heading" a term to indicate that everything was proceeding as it should; that is, that the descent was at the proper rate and the heading was as it should be to bring the aircraft down to and over the end of the runway at its center line. Heilman wondered what the visibility would be like close to the ground, feeling that it should not be so severely restricted as it seemed at altitude. He was right. At the voice command from GCA, "You are over the end of the runway. Take over visually," Heilman looked outside the cockpit and could see the entire NAS Los Alamitos airfield in front of him.

But he also saw something quite unexpected, something that caused him dismay and concern: He found himself in a dilemma because, while his approach had been exactly as directed, he was not over the runway but exactly in the middle between two parallel runways. With but a second to decide before waving off and going around — and unnecessarily going back into the smog and the vectoring of Los Angeles Control Heilman yelled, "Which runway? Left or Right?" Fortunately, Hansen knew and yelled back, "Left!" The quick left turn was easier to make from his left seat in the cockpit than would a quick right turn; and it was made in a descending turn to the runway. Heilman was annoyed with himself for being so obsessed and fascinated with the smog conditions and the area of Los Angeles that he did not focus on whether the left or right runway had been assigned by GCA.

Even though GCA had not positioned them accurately, Heilman was upset more with himself than GCA, feeling that he had really goofed. If his copilot Hansen had not known that the left runway had been the designated runway, there would not have been time to ask GCA or the Control Tower before they would have been too far into the field area to make adjustments and still have enough runway left to land on. And suppose they had been making an emergency landing due to mechanical failure, structural damage, or fuel starvation. Of course, there was no such emergency, but Heilman thought that, had there been such an emergency, his lapse, which created the situation, would have been unforgivable.

Most military airfields have a single runway into the wind — not two, or dual or parallel runways as they are usually referred to. Heilman knew from experience that many commercial airports had dual runways and he was aware that there were some military airfields that had dual runways, so he was particularly annoyed with himself — especially when he had the field chart and instrument letdown procedure page right in the cockpit with him. No doubt, knowing that he would be under the positive control of NAS Los Alamitos' GCA and that GCA would bring him in to the precise point was a factor. He did not have to make an electronic letdown on his own and the letdown page was not as important in this situation. Perhaps this is making an excuse for Heilman, an excuse he would never accept himself. There was a saying taken from a column in *Flying Magazine*, something like, "I Learned About Flying From That." Hanson and Heilman knew that, in addition to your own mistakes, you had to learn from the mistakes of others. But taken from today's lapse, they could categorize it as an experience that they could label "We Learned About Flying From That."

An Instrument Flight Puzzle?

Commander Newton "Newt" White had been flying Naval aircraft for over twenty years and felt he had seen just about everything. He knew, of course, that he had not and could not possibly anticipate — even imagine — all of the things that could happen or all of the challenges that aviators in general and Navy pilots in particular could or would experience. But he had seen much, and very few things that he had seen recently was he seeing for the first time. Still, he was not really prepared for something which occurred on a relatively routine flight in an SP2E Neptune one night as he and his crew of four were flying from an east coast Naval Air Station to the NAS Dallas at Love Field and had just passed Fort Worth. With his copilot LCDR Dan Swensen, he had decided that, since the weather en route was expected to consist of some cloud layers and there was no certainty that they could avoid flying into them, he would file an instrument flight plan. Such a plan would place his flight under the control of the Federal Aviation Administration (FAA), which would naturally assign him a route and an altitude along that route. With such a plan and under the positive control of the FAA, they knew their flight would be smoother since they would probably not have to change altitude or direction to avoid flying through the clouds; they would just fly through them. Although they would still have to watch for aircraft if or when they were between cloud layers, it was very likely that FAA radar would alert them to any other air traffic along the way. Such instrument flight plans were more the rule than the exception, particularly on flights of some distance when it was more than likely that they would have to fly through some weather systems along the way to their destination.

So prepared, Newt, copilot Gus Swensen, crew chief and mechanic Willis, and radio operator Jones — the minimum-sized crew for the Neptune — took off, climbed to the assigned altitude of six thousand feet, and established themselves on the FAA-assigned route. At takeoff, the weather was clear and was not forecast to get too bad. The engines were performing well, as usual, and all the systems were in good order, again as usual. Somewhere west of Pittsburgh, they encountered some scattered clouds; and as they headed southwest through Ohio, Indiana, Illinois, and Missouri in the direction of Oklahoma, they encountered the forecast layers of clouds. The night was very dark, and the air was perfectly calm, eliminating any sense of moving through the sky. The entire crew was wide awake, alert, and yet quite relaxed. There was no need to be otherwise, although they knew that care and attention to detail were advisable on any flight. As they proceeded en route and several hours passed, Newt became aware of something glowing beneath them and felt it came from the lights of some city they were passing over. Although there were cloud layers below them and above them, they did not think that the layers were very dense and such glows from cities are common and were the usual source of such light. He asked his copilot Gus to check the charts to see if the name of the city appeared on the charts. Since the charts were air navigational charts which mainly gave the names of places where electronic navigating radio equipment was located, not every city along the route appeared on these air navigational charts. Gus replied that the city, whatever it was, was probably not big enough to show on the type of maps they were using, and asked Newt if he should get out the geographical map to check further. Thinking that they were probably past the city, Newt felt it was probably not worth the effort to get the map out of the case and told Gus to forget it.

But the glowing light was still there some minutes later when they were decidedly well-past what they thought was the location of the first city; actually it seemed a bit brighter and more yellowish with some blue and green mixed in. By now, Newt and Gus started to feel that the glow was coming from somewhere in or on their Neptune and tried to determine what was causing it. CDR White thought back to the time he first dropped a slowly descending aerial flare one night. The light was so bright and so intense, he first thought the flare had somehow or other ignited and was still attached to the aircraft. But this was different. The light, not intense, seemed to be located just beneath the leading edge of the wing on both sides of the aircraft. It could not have been a searchlight that had been aimed on them; that would have presented a different effect. They were certain it was not a fire and wondered whether it could be one of their own lights that was flickering with a different shade of color because it was loose or about to burn out. But before they could ask the crew members to look out of the portholes to see if they could see anything, the light disappeared; and Newt and Gus were left trying to figure out the mystery. They would, of course, check the aircraft out on the ground at NAS Dallas when they landed, but both were sure they would find nothing. Both Newt and Gus had little to say, but both had been thinking about the strange experience they had had. Some minutes later and at the same time, both exclaimed, "It has to be 'Saint Elmo's Fire!'"

A much mentioned and talked-about phenomenon, every Navy pilot has learned something about Saint Elmo's fire during his training days. It is not something dwelt upon. Indeed, one would be hard pressed to find out much about this phenomenon in books. Even the best encyclopedias contain but little about it. Never in their twenty-some years of flying had either Newt or Gus experienced it, nor had either ever heard of anyone else who had. Sightings of flying saucers were much, much more commonly reported at one time — but not by Naval Aviators. It was reasonable, however, to assume that, depending upon the intensity of the "fire," many pilots may have experienced it without recognizing it. So far as Newt and Gus knew, and they were certain they were right, there was no hazard involved. What it is is best described by a weather professional, a meteorologist like C. Donald Ahrens, who relates it to conditions during which heat lightning is present. From his book *Meteorology Today: An introduction to Weather, Climate, and the Environment* comes this description: "Distant lightning from thunderstorms that is seen but not heard is commonly called <u>heat lightning</u> because it frequently occurs on hot summer nights when the overhead sky is clear. As the light from distant electrical storms is refracted through the atmosphere, air molecules and fine dust scatter the shorter wavelengths of visible light, often causing heat lightning to appear orange to a distant observer.

"As the electric field near the ground increases, a current of positive charge moves up pointed objects, such as antennae and masts of ships.

However, instead of a lightning stroke, a luminous greenish or bluish halo may appear above them as a continuous supply of sparks — a corona discharge — is sent into the air. This electric discharge, which can cause the top of a ship's mast to glow, is known as St. Elmo's Fire, named after the patron saint of sailors. St. Elmo's Fire is also seen around power lines and the wings of aircraft. When St. Elmo's Fire is visible and a thunderstorm is nearby, a lightning flash may occur in the near future, especially if the electric field of the atmosphere is increasing."

What was so impressive to Newt and Gus was that, nearing the ending of their military flying days, they had encountered something from their distant pilot training, something that no one had even spoken about since then. There was absolutely no danger involved, and they really did not experience any thrill or excitement. It was more a matter of, "Well, we have heard about you and almost have forgotten about you, but it's nice to see you really do exist." White and Swenson were somewhat surprised to hear in 1998 a theory proposed that the spark that ignited the German dirigible *Hindenberg* at Lakehurst Naval Air Station in New Jersey came from Saint Elmo's Fire.

Putting the episode in perspective, both Newt and Gus were decidedly underwhelmed and even disappointed that their only perceived encounter with Saint Elmo was disappointing. Any references they had been told about, anything they had read, and any graphic illustrations they had seen, left the impression of brighter, more intense "fire." So much was this their feeling that each wondered if what they had experienced was really the "Fire of Saint Elmo." Based upon the general lack of enthusiasm about the nature of this phenomenon on the part of the public in general, their feeling did not seem far from what was appropriate for their encounter. But even if they could not be certain that they had experienced St. Elmo's Fire, at least they did feel that they had been fortunate for the experience — before they retired.

Part Seven: Mechanical Problems

Mechanical Problems — Surprises and Choices

An experienced Navy lieutenant, taking off one dark night suddenly sees his engine warning light flashing vigorously at him. Is the runway long enough to abort his takeoff? Is fire imminent? What should he do? Another pilot on another dark night senses nothing amiss until he tries to lower and lock his landing gear. The gear will not go down and lock, even though he tries all the recommended procedures. Another pilot takes off from the Naval Air Station at Guantanamo, Cuba, makes a down wind turn to a southerly heading, only to find that, as he heads out over the Caribbean Sea, his engine quits cold. Does he try to get back to the Naval Air Station? Does he make a relatively more safe landing in the sea? And what does a tough young pilot do when, on successive flights, he has first an engine failure that forces him to ditch in the ocean, and then makes a good carrier approach and landing, only to have his arresting hook be yanked from the aircraft, resulting in his being stopped by the barricade?

Should Navy pilots expect the unexpected? Flying has its share of unexpected occurrences; and it may be that, due to the complexity of the state of the art of their aircraft, the military may have more surprises. For perspective, it should be remembered that incidents like those described above are rare, occurring over long periods of time and many hours of flight time. In most, but certainly not all, cases the experienced military pilot has learned not to get too excited; rather, he knows that, with the exception of fire and structural damage, a little patience, common sense, and a reliance on those things he has been taught and learned on the job, he will be able to solve most of those problems he is likely to encounter. He knows further that, if worse comes to worst, he has a reliable means of exiting safely via his ejection seat system.

So, in addition to the many problems Naval Aviators face, and to those mistakes caused by poor judgment, are those unexpected mechanical problems and weather problems which, though not usually resulting in catastrophe, have that potential if they are not handled wisely. Frequently, new aircraft experience problems early on, yet modifications either immediately or in routine overhauls have enabled many of them to have long, productive service lives. The Navy's AD Skyraider series and the Air Force's C-5 Galaxy are two prime examples.

In aviation in general and in military aviation in particular, there is always so much for the pilot to keep in mind, to be aware of, and to be sensitive to. Military aircraft undergo many changes from the time the first models come off the assembly line; two hundred or more changes are not uncommon. Some changes are small and of little consequence; others are large and of much consequence, usually being accomplished during certain overhaul periods. These are normal procedures resulting in an orderly process of bringing about improvements.

The unintended consequence of some of these changes is that the familiar location of instruments and equipment varies somewhat among models of the same aircraft. The P2V or SP2E Neptune was one of the most notorious for the location of various cockpit instruments and switches. But aside from momentary inconvenience, Neptune pilots seemed to experience no real difficulties. On the other hand, some changes in the configuration of the cockpit have presented some inconvenience and even hazardous conditions. For example, a young Navy pilot had quite a few hours in the AD Skyraider, having flown practically all versions from the AD-1 through the AD-4 including the N and Q versions used for specialized missions. In all, he had experience in some ten to twelve different versions. So he felt very comfortable in the airplane, feeling he knew it quite well. But he was in for a surprise, something simple but carrying a potential for catastrophe. Some ten minutes after takeoff on a flight to Norfolk Naval Air Station and at about four thousand feet, he sensed that his flight controls seemed to freeze while he was in level flight. He strained at the controls only to find that he had little effect; although working hard at it, he found he was able to have some slight effect and managed to be able to make a very gradual turn to the left, hoping to return to base. At least, the aircraft remained in level flight — at his present power setting. As he began to think about what the effect of reducing power might be should he return to base, another thought took over. The aircraft behaved as if it was on automatic pilot. His mind raced in considering his situation. No Skyraider he had ever flown had an autopilot in it; yet as he thought, thinking surely this aircraft did not have an autopilot, he looked around at the cockpit side panels, the only place an autopilot could possibly be located. There was no room elsewhere. To his complete surprise, there on the right console were the switches and controls of an autopilot. He immediately pulled up the large engaging knob to disengage the autopilot, and the controls became free again. The electrical "on-off" switch had been left on from the previous flight; and on this flight, the pilot had unknowingly and accidentally placed something on the knob to engage the autopilot. In this incident, the pilot's experience and confidence were both part of the problem and part of the solution.

The potential consequences of this incident may not be readily apparent; but they could have been serious, possibly even fatal. Had the autopilot controls not been set properly for level flight — a distinct possibility — when it was accidentally engaged, the aircraft could have assumed a nose-down attitude, an attitude that could have taken the aircraft into the ground before an inexperienced pilot figured out what was happening. One may easily question how such an incident could occur, but it is easily explained. Aside from the pilot's never having seen an autopilot in an AD Skyraider before, there were several other factors: First, the electrical toggle switch had been left on from the previous flight; next, with so little room inside the cockpit to put things down, the pilot placed his kneepad notebook on the right console, both covering up the autopilot and placing one part of the kneepad in a spot just over the engaging knob; and then early in the flight, the pilot put some weight on the console as he attempted to shift from an uncomfortable position. The engaging knob is a big silver dollar-sized button, which is well-protected from accidental contact by a cylindrical housing having two small finger entry slots for disengaging. In single-piloted carrier aircraft, the cockpit is very confining. The pilot must strap his kneepad just above the knee or perhaps fit it into one of the large pockets in the lower leg portion of his flight suit, the latter being more difficult than just placing it on the right console if he is right-handed. Because this pilot was so embarrassed, it was some time before he could confide his experience to his squadron buddies.

Change, enhancements, and improvements are a necessary part of military aviation, as they are in so many areas of the sciences. This writer was flattered to be consulted as to whether he felt some recommendation to include automatic pilots in the SNB Beechcraft utility airplane should be endorsed. Such a convenience is always to be desired, but its incorporation relative to other improvements has to be weighed, including whether that improvement precludes the installation of other, perhaps more important equipment. If expense is no object and all desired features can be installed, there is no problem; but that is rarely the case, and choices have to be made. Given knowledge of the manner in which rain builds up on the windshield at landing speeds, obscuring forward vision during a crucial part of the flight, and considering the short range of the aircraft, this writer recommended the installation of windshield wipers as a safety feature before the autopilot.

The sophistication of today's maintenance with its computer checks of various systems has done much to provide excellent maintenance of engines and other equipment while the aircraft is on the ground. And in the air in multi-engine aircraft, video displays of engine performance have been around for forty or more years. With the use of this equipment and later improved equipment, it was possible to check each cylinder on every reciprocating engine. Such an airborne check was further assurance to the pilot that, not only were his engine instruments providing valid information, but that the video displays were complementing the instrument information on the pilot's cockpit panel.

Also, among maintenance problems, though thankfully rare, are conditions which arise when good, competent maintenance mechanics make mistakes which become life-threatening — or worse. These mechanics feel terrible, of course, and carry a guilty feeling known only to them — but sometimes to others as well. Undoubtedly, mechanics will make mistakes in the future. Rarely will they be a matter of neglect or carelessness. Most often, mechanics have tried things which, in their best judgment, seem to be logical and reasonable. Naval Aviators complete a form known as the "yellow sheet" after each flight, on which they record the amount of flight time just completed and report any mechanical, electronic, or other discrepancies encountered on that flight. This is a good system, enabling aircraft discrepancies to be taken care of promptly. Occasionally, however, an intermittent problem would be reported, one which, when checked by the maintenance people, would not malfunction. They would then report, "Ground checks okay." This always seemed to be a poor resolution because it could imply that the pilot was mistaken. When, however, several pilots reported the same condition, the mechanics would effect some changes, as, for example replacing parts. Usually, if they failed to resolve the problem quickly, it was not for lack of trying.

Engine Regulators Gone Amuck

The well-known and revered AD-1 Skyraider, like most new models, had its share of problems which arose during intensive use and which would be taken care of through a series of service bulletins and service changes, some to be acted on immediately and others, usually those requiring more extensive work, to be effected during regular overhaul periods. Later models would incorporate these changes and add other features deemed beneficial. And while some of these problems caused anxious moments and challenges for the pilots, which would require perception, skill, and not a little bit of courage on the part of the pilot in dealing with them, fatalities were extremely rare. Of the early Skyraider's problems, an erratic Manifold Pressure Regulator, was one that was not very common but which, nevertheless, occurred often enough to present real challenges to those pilots experiencing it. The MAP worked in conjunction with the throttle, so that, as the throttle was advanced, the pressure and density of the fuel-air mixture (the manifold pressure) delivered to the engine increased, the engine developing an accompanying increase in power. The job of the manifold pressure regulator was essentially to deliver and maintain a constant fuel-air density mixture to the carburetor to insure a constant power setting according to the position of the throttle. And it performed exceptionally reliably — most of the time!

Of the problems associated with the MAP, all of which seemed hazardous, the pilot could immediately be aware of one, while he might not notice the others so quickly.

Some incidents best show the problem. Heading back to Naval Air Station Norfolk, Virginia, with his four plane division following a routine gunnery flight, Bill Spieler was very much surprised, even frightened, as his aircraft engine had a sudden power surge to forty-four inches of manifold pressure, up from the cruising power setting of about thirty-two inches. Realizing he could not bring the power back to cruising power regardless of his position, he immediately increased his engine RPM up to twenty-four hundred to keep from overboosting, and hence damaging, the engine. Too much power for too few RPM's can ruin an engine. Drivers of cars and trucks going down steep declines are advised to downshift, to go into a lower gear to let the engine help to brake the vehicle. Such a procedure in piston-engine aircraft, accomplished by decreasing the RPM causes damage to the engine. The maximum power, ordinarily used only on takeoff is fifty-four inches and twenty-eight hundred RPM. The maximum continuous power setting, referred to as the military power setting is forty-eight inches of manifold pressure and twenty-four hundred RPM. So Spieler's increasing his RPM to twenty-four hundred RPM was good insurance, and it would serve him well should the manifold pressure surge to forty-eight inches. But when normal power settings are from thirty to thirty-five inches and twenty-two hundred RPM, the sudden increase in power brought an accompanying increase in his air speed. The real problem now facing Spieler was how to slow the air speed so that he could letdown and land.

Spieler immediately thought of his dive brakes. They consisted of three large panels which he could deploy to slow down some. Spieler knew that he could lower his speed by raising his nose and, of course, climbing. Such a tactic might be good if he wanted to cut his power and kill off some speed quickly, but then his altitude would increase rapidly, and he would be trading speed for altitude, which would present a different set of problems — but perhaps a more manageable set. In the back of Spieler's mind was the much more important decisions he would have to make about his actual landing approach.

Never having faced anything like his present situation — he wondered if anyone else had had a similar experience and used the brakes, the strong hydraulically-actuated panels, located one on each side of the fuselage and one on the underside of the fuselage. The Skyraider's dive brakes were designed to keep the aircraft's speed in check during dive bombing so that the pull-out from a steep dive could be accomplished more easily and with less loss of altitude. They were not designed like the speed brakes on military jet aircraft or commercial airliners, which could be deployed in increments and which are quite normally used in level flight just to slow the aircraft's speed. The Skyraider's dive brakes were either fully deployed or fully retracted. Spieler was not really concerned that his rapidly accelerating plane was now going too fast to deploy them; accordingly, he did experiment and found he could deploy them and quickly did. But Spieler really wondered if he could handle it. Would he have to bail out, or was it possible to bring the plane in for a landing? He thought he could turn the engine off by moving the throttle to the "idle cut-off" position, thereby starving the engine of fuel.

Another choice was to turn off the magneto switch, which controlled the source of high voltage to the spark plugs. There were risks involved in all of Spieler's choices.

If you shift the scene to mid-ocean and an aircraft carrier in place of a long runway, the situation becomes infinitely more complicated. The mechanics of landing aboard ship involve a very slow, slightly nose high approach, one that would have been practically impossible except for an instant or two while the throttle had been closed and the aircraft was slowing down — and this condition would have to have occurred at a precise time and place or just before receiving the "cut signal," the approval to land aboard. Add to this situation a pitching or rolling carrier deck. Further, add bad weather or reduced visibility.

And in addition, add a black, moonless and starless night. And what about a low fuel state or other mechanical problems? But even in broad daylight, a carrier landing requires that an aircraft coming aboard be in very good operating condition. It is just inconceivable that any aircraft carrier commanding officer would permit what is, in effect, an aircraft with such a fixed throttle control to come aboard: The risk to other aircraft, to the carrier, its personnel, and to the pilot himself would be unacceptable.

Naturally quite concerned, Spieler decided he had better do some concentrated planning. After notifying the control tower and getting clearance for landing, he made a long straight-in approach, estimating how long it would take for the aircraft to decelerate after he cut the engine (i.e. placed the throttle in the idle cut-off position) hoping that he would not have to try to restart his engine by pushing the gas/air mixture forward into the "automatic lean" or to the "full rich" position. Prior to making his actual approach, he did make a practice run to check deceleration after he cut the engine, but did not want to chance a second practice run. In this practice run and making a very careful approach, he cut the engine by pulling the mixture control back to "idle cut-off;" and somewhat to his surprise, he found that he had cut the engine sufficiently far ahead of the runway that his speed, as he approached the edge of the runway, indicated that he was not in danger of overshooting the runway and possibly going off the far end. Since he had planned that approach as a trial run, he did not feel comfortable in making a last second decision to try to set the aircraft down. So going around again for another approach, he was determined to make this his real approach. He did feel, however, that, if he was in danger of overshooting the runway, he felt that he would do what he did not really want to do: move the mixture control forward again to the "auto-lean" or "full-rich" position to obtain power for a third approach. Having rehearsed his procedure, keeping all details in an entirely focused mind, Spieler cut the engine far from the runway, the aircraft slowing down so that, if far from perfect, he could restart his engine and go around for another try. But Spieler brought the Skyraider down to the runway in a very strange sort of "dead stick" or power-off landing, certainly one that, if not perfect, was not too far from it — and one that few pilots have ever had to do.

His squadron buddies were impressed with both his cool handling of a difficult emergency, and with his managing the aircraft. But however it looked to others, to Spieler, it was not just something routine; and he was not above saying that he did not ever wish to have to confront that situation again. It did not seem at that time that much was made of these incidents, perhaps due to the relative infrequency of the occurrences. So there was very little in the way of a discussion, nor was any company representative brought in to try to explain to the pilots this aberration. Both an explanation and a review of the best way to deal with situations involving erratic Manifold Pressure Regulators might have been helpful and reassuring to all the pilots, particularly for those having been involved in dealing with the problem. But, for the most part, the pilots did not seem especially concerned or even particularly curious, no doubt feeling that Spieler's situation was a one-time occurrence. Even those who were interested to some degree in the details of how Spieler had handled his problem so well never had the opportunity to learn directly from him. LTJG Spielman was killed shortly later bailing out of an AD Skyraider, while on a training mission.

LTJG Bud Harmon of the same squadron experienced another problem involving a Manifold Pressure Regulator, again in an AD Skyraider, but an incident of a different sort and occurring, during takeoff when the aircraft was not yet airborne. The weather was clear and the stars and moon were bright as Harmon was taking off on a night training flight out of Naval Air Station Norfolk. Harmon, in following standard procedure which called for the pilot to taxi out to the warm-up area next to the runway, advanced the throttle to thirty inches of manifold pressure with the propeller control set to "full low pitch" or the highest RPM setting, and then checked the magnetos. Harmon, having done this, and having received acceptable RPM drop-offs as first the left bank of spark plugs was checked, then the right bank, and finally both banks at once, felt confident in the engine's performance. After making the established preflight checks of the engine, and in accordance with established procedures, Harmon received takeoff clearance from the tower, taxied out on the runway, where he advanced the throttle to thirty inches of manifold pressure with the propeller control in the "full low pitch" setting. At thirty inches and again according to procedures, which were designed to keep the nose from tipping over and damaging the propeller and engine as more than thirty inches of power were generated, Harmon released his brakes and advanced the throttle to the full power position which should provide about fifty-two-to fifty-four inches of manifold pressure. LTJG Harmon had gone through this procedure with the AD-1 and had taken off over a hundred times. But this time was much different.

It is a tribute to this marvellous aircraft, which was known to be capable of carrying its own weight in payload, that LTJG Bud Harmon, and perhaps many other young pilots as well, would not immediately recognize his condition. Perhaps because the aircraft had no external fuel tanks and he was carrying no payload the aircraft at first seemed to accelerate fairly quickly. It was not until some seconds later that Harmon felt his acceleration was not what it should be. Was he mistaken? Did it just seem that way? Did the wind suddenly increase? Harmon quickly checked his throttle position to make sure it had not slipped back and then checked his engine gauges. His throttle setting was okay and his airspeed was building up. But wait! His manifold pressure indicated <u>forty inches</u> not the fifty-two to fifty-four it should have been. But what to do? He was well past the runway halfway point and was just about airborne. He could still abort safely he felt — provided he could close the engine down quickly — but then again, why abort? He knew his aircraft was light because he had neither fuel in his external tanks nor ordnance on the racks. Now airborne, he knew he could climb with the power he had; and besides, sometimes aborting brought problems of its own: Trying to brake, he could have run off the end of the runway or have braked too hard and caused a "nose-up" — the nose to tip down and ruin the propeller and engine, perhaps even flip over on its back, not an unheard of occurrence. In fact, as Harmon appraised his situation, he really found nothing to be concerned about. He found this experience very interesting, even intriguing, and a little exciting without being dangerous. In flight he had, of course, more power than he needed, thirty inches being the normal cruising power setting.

Harmon berated himself for not noticing immediately that he could only get 40 inches of manifold pressure immediately; but since the design of the aircraft and its roughly three thousand horsepower engine require that the aircraft be rolling down the runway as power beyond thirty inches is employed, realization of his inability to get full power did not and could not occur immediately. With aircraft having a tail wheel and not a nosewheel, it is not difficult to nose up the plane when the brakes are on and power is advanced beyond a certain point. That said, however, Harmon should have monitored the manifold pressure and the RPM as they built up during the first part of the takeoff run. But how many times had he routinely advanced power without any hint of a problem?

In his nonthreatening current situation, Harmon started to think about the details of his takeoff, wondering if he had really observed things accurately? Had the throttle been stuck at some intermediate point where, with just a bit more forward pressure, the throttle would go to full power? The more he thought about things, the more he wondered. Somewhat annoyed with himself for his carelessness and lack of observation, he decided that

Mechanical Problems 277

he had to find out. Feeling that he did not have an emergency situation, he decided to land and then take off again, this time being certain to check everything carefully. How many pilots would have made this decision would be hard to know. Surely some, perhaps most, would be critical of such a decision. Others might have decided as Harmon. Harmon wondered what the Navy test pilots at Patuxent River would have done. He liked to think that their decisions might be similar to his.

Receiving takeoff clearance for his second takeoff, Harmon released his brakes as he advanced his throttle beyond thirty inches, careful this time to see if the throttle moved the full length of its travel to full power. Sure that the throttle was fully advanced, he then looked to see whether the manifold pressure reading was only 40 inches. Knowing that he had full-throw of the throttle, he expected that the manifold pressure might creep up further — but to no avail! Harmon continued his second takeoff, one exactly like the first — only this time he knew what he had to look forward to and how he could describe accurately to the maintenance people what had to be fixed.

The Wright Radial 3350 engine mounted on the AD Skyraider was an excellent, powerful engine, one, which as previously mentioned would let the AD carry its own weight in ordnance or in a combination of ordnance and extra fuel in its external, droppable tanks. Certainly for most operations with heavy fuel and ordnance loads, unlike Harmon's easy night flight, all of the engine's power is needed. Aboard ship, for example, whether for catapult launches or normal deck takeoffs, the conditions Harmon experienced could have had very serious consequences. How Harmon would have handled the limit of forty inches of manifold pressure on a takeoff from an aircraft carrier with external fuel tanks and a load of ordnance would be quite another situation — one which Harmon chose not to think about.

The Steady, Reliable Avenger

In a quite different situation, was an incident involving a TBM Avenger, a torpedo plane which had been taken out of moth balls or storage. The Avenger was an old aircraft, one that had been used in World War II, mostly in the Pacific Fleet. It was slow then, so its use was restricted as much as possible to certain combat situations. Over the years, many Navy aircraft, after receiving certain pre-storage preparations, had been stored locally in steel containers or flown to the desert at Litchfield Park, Arizona, where they were literally parked in desert-like conditions in the open air, a storage which, due to the dry desert air, inhibited deterioration. How often these aircraft have been taken out of storage to meet needs is not known; however, there was one significant period of time when TBM's were removed from storage to meet an immediate need. In the late 1940's and early 1950's, when the size of the Soviet submarine threat was perceived to be alarming, there were really not enough adequate carrier aircraft specifically designed for anti-submarine detection and warfare. The newer carrier antisubmarine aircraft, the cumbersome Grumann TF Guardian, would be used in the Korean War and would be superceded by another antisubmarine aircraft, the four-man Grumann S2F Tracker. In the meantime, the "hunter-killer" concept of using two aircraft — one a plane with detection equipment and the other an aircraft with weapons to destroy the submarine — came into being. Not really the best aircraft for the Soviet submarine threat, the TBM was, nevertheless, a good stable, reliable, adaptable aircraft — and there were many immediately available from storage — if only to meet a temporary need.

With the Soviet submarine force of well over four hundred and growing, the threat was real; and the need for an enlarged antisubmarine capability was urgent. So bringing the TBM Avengers out of storage was the means of enlarging the capability quickly. The aircraft so retrieved from storage performed very well, and the decision could well have been considered wise, since it performed so reliably for the intended purpose. It is the exceptions to the rule, however, which prove interesting, one of which resulted in an unexpected challenge for a young Naval Aviator.

Returning to the NAS Norfolk in his TBM Avenger at the completion of a night mission, LTJG Bill Higgins was surprised to find that, when he placed his landing gear lever in the "down" position, the wheels moved out of the "up and locked" position but stopped somewhere between the "full down" and "full up" positions." The situation, of course, required that Higgins abort his landing approach and notify the control tower personnel that he had a problem with his landing gear. He then placed the lever in the "up" position and the wheels were retracted by the hydraulic system to the "full up" position. Although Higgins was not about to concede that he might have to land wheels up and all that that could mean, he had to explain his situation more fully to the tower and then try to reason what the problem could be and what he might possibly be able to do about it. He told the tower he would fly clear of the air station to an area of less air traffic to experiment and was not sure how long that would take. His first thoughts were reassuring: He had plenty of gas, at least about two hours worth, and the weather was good and forecast to remain so, removing two potentially complicating factors and allowing him time for a thorough assessment while reducing the risk of a midair collision. He wanted to know whether he had a general hydraulic problem affecting all hydraulically actuated systems or whether his problem affected only his landing gear system. Slowing his aircraft, he placed his flap lever in the "down" position and was pleased to see that that system operated properly, removing the need to have a higher landing approach speed should he have to land wheels and flaps "up." A higher approach speed in a wheels up condition would increase the length of his slide on the runway, creating more heat, sparks, and even spilled gasoline and oil and all that that could entail. Higgins was satisfied the flaps were okay and then checked his brake system by depressing his pedals. Being pleased with the feel of the pedals, he knew he could count on having some braking provided he could get his wheels down.

Now, Higgins could concentrate on getting the wheels down. As designed, when being raised, the wheels on the TBM Avenger moved out away from the fuselage at right angles and toward the wing tips. There was no forward or aft movement, so that, when fully retracted, they lay flush along the underside of the wing and parallel to the leading edge of the wing, with the actual wheels and tires in the direction of the wing tips. Nor was there any turning or twisting of the landing gear as there would have been if the landing gear retracted aft and the wheels had to rotate ninety degrees to lie flush with the wing. Other airplanes had similar design features, but most of them retracted with the wheels moving inbound or toward the fuselage. Higgins knew that the TBM design would be helpful if he could not get the wheels down hydraulically.

He again put the landing gear lever in the down position, hoping that this time, whatever the previous problems, the hydraulic system would do the job. No luck! The wheels came part way down and stayed there.

Earlier, when Higgins had called the control tower to tell them why he could not land, the tower personnel in turn called Higgins' squadron duty officer to relay information about his situation. Apparently within minutes, the squadron skipper and maintenance officer were on the way to the tower. As Higgins planned his next step, the skipper's voice came on the air asking Higgins to describe his difficulties. After explaining his situation, the skipper suggested what Higgins should do next — exactly what Higgins himself had determined would be his next step: He was advised that with the landing gear lever in the down position and the landing gear in the part-way-down status, he was to rock the wings sharply so that the forces so induced would jerk one wheel and then the other into the full-down position. If the maneuver worked as planned, the sharp upward movement of the left wing would tend to move the left landing gear toward the full-down position. A similar sharp upward movement of the right wing would have the same effect on the left wing. But again and even after numerous additional attempts, the situation remained unchanged.

While realizing that he had a couple of avenues yet to try, Higgins thought that perhaps he might have to land wheels-up, an action he really did not want to have to take. Returning an aircraft to base with its wheels down, even if you had to allow for certain hydraulic problems, was worlds better than landing wheels-up, damaging the propeller, engine, fuselage, and who knows what else. Additionally, although he realized that it sounded foolish, his pride and determination were involved. In considering what courses of action remained, Higgins considered two. First, with the wheel lever in the down position, he could dive to pick up speed and then pull up abruptly pulling several G's (gravity forces) on the airplane including the partially extended wheels, hoping that this force would move the wheels toward the full-down position. He was really doubtful that, if the abrupt rocking of the wings, which had exerted some G forces also, had not produced the desired effect, additional G forces would work either.

Throughout these efforts and constantly in the back of his mind, was the thought that, since the problem seemed to lie solely with the hydraulics of the landing gear system, operating that system in the usual manner would probably just not work. At this point, Higgins received another call from his commanding officer asking him how he was doing and what he planned to do next. Higgins sensed by the skipper's tone of voice that he wanted to check on Higgins state of mind and that he too was now considering the possibility of a wheels-up landing. Higgins reported that he was fine and that he wanted to try something that just might work; he asked the skipper what the maintenance officer thought of the idea. To get around the faulty operation of the landing gear hydraulics, Higgins wanted to take the landing gear lever out of the full-up position; but instead of moving the lever to the full-down position, he would leave the lever in some intermediate position. The maintenance officer had been mulling over several theories himself and thought that Higgins' idea might just possibly work. The condition, after all, was not an emergency covered in the aircraft's handbook of emergency procedures, so other possibilities were called for.

Checking the airspace around him for clearance, Higgins placed the landing gear lever in an in-between position to see what would happen. As expected, the wheels came part way down — to the same position as in earlier attempts. Now, Higgins would rock the wings vigorously to check his theory. While what happened next was not really a surprise, the speed with which it occurred did surprise him: Both wheels snapped promptly into the "full-down" position. There was relief all around, but Higgins was still cautious.

He had been around long enough to know that problems people thought that they had solved had a way of cropping up again. Although several questions came to mind, the immediate problem dealt with the intermediate position of the landing gear lever. Should he land with the lever in that in-between position or should he now put it in the "full-down" position? Just as Higgins completed that thought, the maintenance officer's voice came over the radio instructing Higgins that now he should place the lever in the "full-down" position. While Higgins was in agreement, he wondered first if the lever could now actually be placed into the full-down position, and given the strange operation of the system so far, if somehow the wheels would come partially up again, causing for some unknown reason a condition in which he could not get the wheels down again. Nevertheless, Higgins felt the advice he received to be good advice, perhaps because he too felt that it was the most logical place for the lever to be; so he moved the lever from the intermediate position toward the "full-down" position, being relieved and a little surprised that the lever moved into that desired position without any resistance. The 'full-down" position usually included a locking pin for additional safety: and Higgins felt that, if the wheels were down and the landing gear lever was in the "down" position, it was very likely that that locking pin would be in place. As he really expected, the wheels remained in the full-down position, and Higgins felt relatively confident that the decision to put the lever in the "full-down" position was a good one. If the lever had been left in an intermediate position after Higgins landed and, for some unknown but unlikely reason, the landing gear system began operating in its normal fashion, the wheels would come up or collapse on the runway, resulting in the very thing they had all worked so hard to avoid. Many aircraft had micro switches in the landing gear system, so that, once on the ground with the weight of the aircraft on the wheels, it would be impossible to raise the wheels.

With later aircraft, most if not all, had these micro switches, so that, if the pilot accidentally raised the landing gear lever instead of the flap lever after landing, he would be protected against his own mistake. But the TBM Avenger, at least some of them, did not have these micro switches. It was possible to raise the wheels while the aircraft was on the ground; and this was the concern of everyone about what could possibly happen if Higgins had not placed the lever in the full-down position.

When Higgins was ready to land, he felt quite good about the condition of his aircraft, although he could not really be absolutely certain. Cleared to land and touching down some seconds later, Higgins felt that the roll of the aircraft on the ground was just right and was now certain that his wheels would not collapse. At some subsequent point during his landing rollout, Higgins could not later recall exactly when, he just forgot about the entire problem and taxied to his squadron tie-down spot. He felt no need for a cold, refreshing and relaxing beer — a much more stressful situation would be required to reach that point.

In working through what Higgins thought was the cause of his problem, the idea occurred to him that the problem was such that the hydraulic piston, which pushed the wheels into the down position or raised them, had some basic and simple flaw — perhaps a leak. Higgins figured that, if all other hydraulic systems worked (flaps, bomb bay doors, etcetera), there was some malfunction in the landing gear system alone which precluded the full extension of the landing gear to the down position. While he could not be positive, he thought that there must be hydraulic fluid getting on the wrong side, or both sides, of the hydraulic piston which lowered the landing gear and that the piston could not compress it, thereby leaving the wheels in an intermediate position. While it was just a theory to Higgins, he was not exactly surprised to learn a day or so later that his theory was essentially correct. A neoprene seal had dried out while the aircraft was in storage, and the hydraulic pressure against the seal created cracks allowing fluid to get where it did not belong. How many situations like this occurred in the TBM Avenger Higgins never learned, but it was reasonable to assume that there were other similar incidents — and if not, certainly the potential for others existed.

The Pros Don't Quit

Pilots involved in flying demanding military or commercial aircraft often face life-threatening risks; sometimes these risks result in accidents with deadly consequences. Pilot error is one factor which every pilot seeks to avoid. But not all accidents are caused by pilot error. Mechanical failures, faulty maintenance, and weather conditions which the pilot cannot avoid are sources of problems leading to accidents. And despite the use of modern satellite weather information and improved weather forecasting, not all conditions can be known or predicted accurately. But, so far as problems arising from mechanical failure or faulty maintenance, the commercial airline pilot has the advantage: he belongs to a union, making it much easier for him to raise questions about "safety of flight" considerations than the military pilot who has no such avenue. But that is not to say the Navy's safety procedures are lacking in any way or that very serious efforts are not made to correct reported discrepancies.

Several incidents serve to describe both mechanical failures and faulty maintenance which can occur and also the degrees of seriousness which can result. First, there is the case of feisty Dick Allen, a short hot-tempered lieutenant junior grade with guts and competence. Nobody messed with Dick, regardless of size; yet everyone liked and admired him for his hard working, uncomplaining nature and human qualities. Dick had taken off from the aircraft carrier *USS Coral Sea* (CVB-43) in his AD-1 Skyraider to rendezvous with his four-plane division for some simulated bombing runs. Shortly after taking off, and at an altitude of only eleven hundred feet, for some unknown and never-to-be-determined cause, his engine quit cold. Feeling he would probably be too low to bail out; and after quickly assessing his situation, he realized that he faced two problems: he didn't have enough altitude to turn his aircraft into the wind for ditching in the ocean; and there were big wave swells on the water with correspondingly deep troughs between the crests of the swells. Even if he had had enough altitude to turn into the wind, taking a chance on striking one of the crests did not seem very desirable. That left landing in one of the troughs his best option — and even that, given the very brief time he had, would be difficult to accomplish as he would wish. But manage he did, doing the best he had time for to make minor heading corrections to land parallel to the crests in one of the troughs. Landing in one of the troughs meant that he would be landing at right angles to the wind, or a condition like making a landing in no-wind conditions; so his speed on hitting the water would be high, and the impact would seem more severe than he had anticipated. Additionally, to compensate for the crosswind, Dick had to head slightly into the wind (crab) or keep one wing slightly low (slip), configurations either of which would make his impact with the water less than ideal. But as it turned out, other than the unexpected jolt, he was okay and had no trouble getting out of the Skyraider and into his "Mae West" life preserver and then into his life raft. A rescue helicopter, which had been airborne and "on station" for just such emergencies, was over him in what seemed like seconds. And he was back aboard ship in good condition in minutes. Other than to say that he was really surprised that he had hit so hard because he could not land into the wind, he had little to relate; and his nature would not allow him to be fussed over.

The Skyraider turned out to be a wonderful workhorse of an airplane. It could easily carry its own weight in rockets and other ordnance; and it earned the affectionate name of "The Spad" after a revered World War I aircraft. But the AD-1 model of the Skyraider, as the first of a long series of Skyraiders, went through a long period during which there were some so-called engine failures. The word or explanation which eventually came out was that it was not the engine, the very reliable Wright Radial R-3350, but something very unusual. The explanation received was that the problem was in the fuel tank, the huge self-sealing fuel tank located just in back of the cockpit had internal baffles made of similar material which were designed to keep the gasoline from sloshing about during maneuvers, particularly when the tank was less than full. According to the report, sometimes these baffles would break loose and would cover the port from which fuel was drawn from the tank for the engine. The explanation seemed plausible, particularly since the history of the Wright R-3350 would prove it to be a marvelously reliable engine.

But to Dick Allen, all that did not seem to matter. His nature would not allow him to be consoled or catered to. He expected and wanted to get right out there and fly again. So, the next day, he donned his flight suit in the ready room, received his flight briefing, and went up to the flight deck on the command of "Pilots man your airplanes!" and took off on another mission, forgetting the previous day's events. This flight was completely routine, going exactly as planned, right up to the divisions' flight back to the vicinity of the ship. There would have been no way Dick, or anyone else for that matter, could have anticipated what would happen to him at the end of this flight. Taking his place in the strict landing order, Dick made his approach to the carrier when his turn came; and being a very able pilot, his approach to the stern of the carrier was all that could be desired. The "landing signal officer" gave him a "cut signal," and Dick cut his engine to ease his aircraft down onto the deck and to have his tail hook catch one of the arresting wires stretched across the width of the deck and bring him to an intended abrupt but safe stop. Dick felt the deck under him and had sensed some contact with an arresting wire. But when he did not feel the solid feel of having his tail hook catch a wire and when he saw himself going up the deck past about six or seven arresting cables, he naturally wondered why. With the aircraft moving up the short distance of the deck towards the barricades, there was precious little time to wonder what was happening. Within seconds, his aircraft contacted the barricade cables, two strong steel cables of about one inch in diameter strung across the deck at heights of about four feet and seven feet. His aircraft on contacting the lower cable nosed up so that the propeller blade tips were bent back and folded back a bit over the engine nacelle. Fortunately, the Skyraider did not flip over on its back as sometimes happened. Dick was not hurt, but he was angry as hell. Climbing out and down from his aircraft, he yelled quite forcefully, "God damn it! I quit!" This expression was to be uttered repeatedly as he walked toward the island superstructure. There he was met first by the air group's flight surgeon, whom everyone knew, and who calmly approached Dick, put his arm around his shoulders, and ushered him into the medical spaces in the superstructure island located at flight deck level. In those medical spaces, the doctor broke out one of the little two or three ounce bottles of brandy kept for special circumstances and gave it to Dick, who unconsciously consumed it. That Dick would let anyone console him so was a surprise to everyone. But it was generally assumed that, in his angered state, he did not realize that he had given into such pampering. Whatever else the flight surgeon did, Dick, calming down in rapid order now, shortly made his way down to the ready room to take off his flight suit, put on his uniform, and, hoping to find some explanation for the accident, try to place everything in perspective. No one approached him to ask questions; rather, everyone who knew and respected Dick gave him his distance, knowing full-well that he would talk when he was ready.

Dick really did not think that the accident was caused by anything he had done and thought the most likely cause of the accident was what was known as a "hook bounce or skip," a situation which causes the hook to bounce up and down after first striking the deck, often bouncing so that the hook strikes the deck just between the arresting wires as the aircraft moves up the deck. Hooks are normally kept from bouncing by some hydraulic dampening system, a system obviously malfunctioning when hook bounces occurred.

Dick and every other Naval Aviator had known of hook bounces; and he had also known about other hook problems. Thinking his accident had been caused by a bouncing hook, he was surprised to learn that his hook had actually engaged an arresting wire but that, as soon as it did, the hook was pulled right out of the aircraft and lay on the flight deck. Other pilots felt that somehow this knowledge, however it might have upset him coming the day after his engine failure, provided a measure of relief to him.

Although today, hook bounces are rare and hooks that come off are even rarer, the barricades are still necessary; and in Dick's incident, the barricades did exactly what they were designed to do. At least his aircraft did not flip over on its back after nosing up as was so frequently the case. In such a situation, there is the problem of getting back down on the deck after releasing the seat belt. More than one pilot has released his seat belt while hanging upside down only to bang his head on the flight deck. No doubt, the reader wonders whether Dick's saying, "God damn it! I quit!" was serious. But for the "feisty Dick Allen," unsurprisingly it was something that he did not remember saying. Even if frustrated enough to say it, his inclination without question, would have been to control his behavior and avoid such an outburst — there was no one who would argue with Dick that he ever said it! What would be the point? Needless to say, Dick Allen did not quit, he did not hand in his wings, and he did fly the next day, wondering what perhaps new and unexpected event would occur. Of course, the next flight and the next and the next *ad infinitum* were perfectly routine. And feisty Dick continued a long, successful career of flying as you can well imagine.

A Life or Death Situation — What Would You Do?

For the uninitiated, a "Hangar Queen" is an aircraft which is usually out of commission for one reason or another, often for a long time with the result being that it is not flown very much. Sometimes, there is damage which can only be repaired at overhaul facilities. Sometimes, the aircraft needs a part or parts which are not immediately available or must be sent for far from the aircraft carrier at sea. Still others may exhibit problems which seem to defy adequate and long-term solutions, the consequence being that men and facilities must be directed elsewhere until there is time to focus on that troubling discrepancy.

Universally condemned in the military but nevertheless resorted to is the removal of parts from a seldom-flown aircraft, the hangar queen, and the installation of those parts in another aircraft. The process is referred to as "Cannibalizing." In the year 2000, a prominent example of this was the removal of thrust reversers from an engine nacelle of an Air France "Concorde" and its installation on the nacelle of another "Concorde." This cannibalizing would hardly have made news were it not for the crash of the receiving Concorde, and Air France wanted to assure everyone that the cannibalized thrust reversers had nothing to do with the crash. In the military, commanding officers are often faced with difficult choices, making them sometimes authorize the taking of a part or parts from a hangar queen. Sometimes, maintenance officers have but little choice and so engage in the process. But it is when so many parts have been cannibalized, that the process has skeletonized the aircraft, that the result is very unsatisfactory.

Most Navy pilots have known hangar queens of one description or another and the process known as cannibalization. Maintenance officers could remember many such aircraft. Descriptions of these queens, either skeletonized or just awaiting parts or a diagnosis of some reported discrepancy, are not very interesting unless there is some follow-up related incident. Naval Aviators will have their own memories of some of these so-called queens. It would serve but little purpose to go on and on about them, but some information may put perspective on the kinds of things that can happen. Not certain but often enough, incidents occur to aircraft when first flown after being out of commission with one thing after another. Earlier, an incident was described concerning an F8F Bearcat which was brought up from the hangar deck and launched with unexpectedly tragic results. On a Mediterranean Sea tour, an F4U Corsair had a series of problems; and on one of the flights from the aircraft carrier *USS Midway* (CVB-41), engine trouble caused the pilot to make an emergency landing at the now-closed Wheelus Air Force Base in Libya. A team of mechanics was sent to the air field to work on the engine; and when the F4U was deemed ready to fly back to the ship, the squadron's maintenance officer wanted to fly the F4U Corsair back aboard the *USS Midway*. After giving the aircraft a good preflight check and feeling that the engine's troubles had been resolved, he obtained clearance for takeoff. The aircraft climbed just a few hundred feet before the engine failed, and the pilot had to make a wheels up landing just outside Wheelus AFB in an arid desert-like area. How long it remained there is not known. No doubt the Air Force retrieved it, but it was of no further use to the squadron or the Navy.

The previously referred to "reliable TBM Avenger" was taken out of "mothballs" to help in the frantic attempt to contain the expanding Soviet threat of four hundred or more submarines some four or so years after World War II. It was a bit after Winston Churchill's Fulton, Missouri, speech in which he put forth the famous words: "From Stettin in the Baltic to Trieste in the Adriatic, an Iron Curtain has descended across the continent. Behind that line, lie all the capitals of the ancient states of central and eastern Europe: Warsaw, Berlin, Prague, Vienna, Budapest, Belgrade, Bucharest, and Sofia; all of these famous cities, and the populations around them, lie in what I might call the Soviet Sphere, and all are subject, in one form or another, not only to Soviet influence but to very high and in some cases increasing measure of control from Moscow." By the time the TBM Avengers were brought out of their states of preservation, the cold war, and all that that entailed, was firmly established. To some extent, it may have been the rush to make the Avengers available as quickly as possible that caused some of the problems; but, all in all, it is felt that the problems were relatively few. Of these problems, one stands out, one potentially much more serious than the one described earlier in this chapter. The problem was exacerbated by the trying conditions the pilot had to face and the choices the pilot was forced to make.

The reader may wish to place himself in the cockpit of this airplane to try to see what the pilot faced and what, perhaps, the reader may have wished to do in this instance. A squadron of antisubmarine Avengers had been training at the Naval Base at Guantanamo Bay, Cuba, and McCalla Field, its Naval Air Station, in preparation for doing some night aircraft carrier qualification work, including night carrier landings. Concerning one particular Avenger, several pilots had reported, after flying the aircraft, that the engine was running rough or was "missing," as the saying goes, to indicate that it seemed that one cylinder or a spark plug was not functioning properly. Some even thought it could be preignition or "knocking" or perhaps a problem with the ignition cable; but these were less widely held views. After each of these flights, the mechanics would check the engine and make adjustments or replace parts as they felt necessary or warranted. But still the pilots' reports were the same. How dangerous the condition was no one really seemed to know. But ordinarily such a condition would be remedied routinely; and since it almost always was, the pilots felt that there was certainly no point in flying an aircraft with a misfiring engine. From the number of reports and the aircraft's being in a "down" status for some time, it was acquiring the notorious appellation of "Hangar Queen." With many pilots concurring, the condition seemed most likely to be an ignition problem, one which a faulty spark plug which was not firing or not firing at the proper time could cause. None of the pilots thought it was a fuel problem, so they were surprised to learn that the mechanics had decided to change the carburetor, a very sizable device fifty or more times the size of an automobile carburetor. Having tried so many things up to this point and having gone to all the trouble and effort to change the carburetor, the mechanics felt confident that they had finally corrected the problem, one mechanic even stating confidently that they had finally fixed the problem. Though surprised, no one questioned the decision, feeling that the mechanics were probably right.

Assigned to the first flight of the Avenger after the carburetor change LTJG Buzz Windsor checked the engine thoroughly before deciding to take off. With everything checking out normally, Windsor and his crew of two took off on the north runway. He was to turn south out over Guantanamo Bay and then farther south over the Caribbean to rendezvous with three other aircraft for their exercise. Buzz made his climbing left turn out over Guantanamo Bay and headed south. As he turned downwind parallel to the runway he had just taken off from, he continued to climb, reaching 1,800 feet when he was just opposite the point at which his takeoff run had begun. At that point, and with no warning of roughness, the engine quit cold, creating an awful silence and an engine that could not to be started again. Having a dead engine in a single-engine aircraft is very bad news. It's a condition know as a "dead stick." It is never good news to lose an engine in a multi-engine aircraft; but in a single engine aircraft the pilot knows he's going down and must look for a landing spot. But Buzz' position was not too much different from that of a pilot who was in position to start his normal landing approach by making a descending left one hundred eighty degree turn to line up with the center line of the runway for landing. Bailing out was an option. But why should he not be able to make a landing approach similar to a normal approach and landing. After all, at 1,800 feet, he was higher than the normal height to start the approach; and even if he had no engine, he would have enough altitude to reach, not undershoot, the runway.

To appreciate the situation Buzz Windsor faced, some further information about McCalla Field is helpful. The Naval Base at Guantanamo Bay is on territory leased from Cuba as part of events following the end of the Spanish-American War in 1898. The agreement, signed in 1902, permitted the United States to use the area for a period of about ninety-nine years and would have expired, without extension, in the year 2002. There have been two airfields there for some time, the first being McCalla Field on the east side of Guantanamo Bay, which was established in the early days of Naval Aviation, and the second being Leeward Point across Guantanamo Bay to the west. Established more recently, Leeward Point has much longer runways to accommodate jet aircraft. McCalla Field is part of the main complex which extends to the east, and as such is very close to and convenient to all of the main installations. One might wonder then why the runways at McCalla Field were not lengthened to permit safe jet aircraft operations as they had been at practically all of the Naval Air Stations in the United States. A visit to the field or perhaps a photograph or two taken from the right perspective would reveal just why lengthening the runways was not feasible or might even be impossible. There are only two runways at McCalla: one is a North-South Runway, and the other is Northwest-Southeast. The field is at an elevation of about eighty-five feet above sea level and is bordered on the south and west by sheer cliffs from the runway level down to the sea. On the north, the situation is not much better. The runway drops off abruptly dropping down almost the same distance to a seaplane area paved with concrete. The Northwest-Southeast Runway is very short, the runway to the southeast again coming in over the cliff on the west side of the field and the drop-off at the runway end being less sheer and not as deep. Indeed, for some Naval Aviators, and others as well, their first visit to McCalla Field would be at a time when an aircraft which could not stop in time lay at the bottom of the drop-off at the southeast end of the runway, where it might remain for awhile before it could be hoisted and carried away.

Landing at airfields where the approaches to runways include cliffs is not a problem, nor is it a mental hazard. It is like landing at any other field and like landing on a plateau. Pilots who have essentially only flown jet aircraft with their flatter approaches, higher landing speeds, and much longer runways are less likely to appreciate some of the kinds of approaches to landings made in aircraft powered by piston engines with their propellers. The Northwest-Southeast Runway at McCalla was only 1800 feet long — and seemed shorter — requiring, to begin with, that the wind be very strong and right down the runway and that the pilot plop the aircraft at the very beginning of the runway. That runway at McCalla has long been closed to fixed wing aircraft, as are many similar though longer runways at Naval Air Stations. The Naval Air Station Anacostia near Washington, DC, now closed in favor of Naval Air Station Andrews in nearby Maryland, had a short East-West Runway leading to the Potomac River on the west. A pilot landing in this direction had to consider not only the short runway but also some hazards at the approach end, where he had to just clear or get over the crest of a hill, avoid some nearby menacing smokestacks, and drop down sharply if he wanted to touch down on the runway at its beginning. Failure to do this precisely would cause the pilot to land farther down the runway and perhaps wind up in the Anacostia River. The airfield at the Naval Air Development Center Johnsville in Warminster, Pennsylvania, had a short, challenging runway. It was about 2,200 feet; and depending on which end the pilot approached, he could also be landing slightly downhill — and severe braking in aircraft with tail wheels (and not nosewheels) was not advisable because the aircraft could tip up damaging the propellers — and worse. Now, these types of runways are not in use; and, in fact, many if not most of these fields are closed. But sometimes Naval Aviators who are required to fly into small civilian fields encounter unexpected or unanticipated approach challenges: for example, certain approaches at civilian fields in Lynchburg, Virginia, and Harrisburg, Pennsylvania, provided unusual requirements for military aircraft. At Harrisburg with certain wind conditions, the pilot faced a short runway, which seemed to begin just a few feet beyond some tall trees, requiring that the pilot just clear the trees, then cut back engine power abruptly, and drop down quickly to the edge of the runway. The short military runways still in use when these fields or particularly short runways were in use were really anachronistic, dating back to much slower, much smaller, less sophisticated aircraft and times. And for obvious reasons, military pilots in their modern aircraft are well-advised to consider carefully which small civilian air fields they decide to fly into.

It may seem incongruous for Naval Aviators, who land on aircraft carriers in spaces between 200 and 400 feet to even mention short downhill runways with obstacles to clear. But consider that, when Navy jet aircraft first became operational, pilots were landing them on 5,000 foot runways, runways whose length required similar approaches to utilize the full length of the runway — and even then, arresting cables were frequently installed at the runway ends to prevent running off the end of the runway.

Buzz Windsor knew McCalla Field well. He knew the runways and what would be required, so he decided to try the approach, knowing he had some leeway in starting, at 1,800 feet, a bit higher than the normal approach. He knew the complicating factors presented at McCalla Field. He knew that, of the two runways, there was only one realistic choice — landing on the runway he had just taken off from and heading in the same takeoff direction. This would give him 5,000 feet of runway and allow him to land directly into the wind. Still his position was not bad — at least he should be able to reach the runway. But the further and more imposing complication remained: the field's elevation above the sea and the steep rock wall or cliff extending on both the approach from the south end of the runway and on the west side of the runway. Not only that but there was not much of a grass area between the edge of the cliff and the beginning of the runway, a factor which meant that, in effect, Windsor had no safe extension of the runway in case of an error which had him landing short. Running through his mind also was his knowing that the other end of the runway was equally bad, with its equally bad steep drop to the paved concrete seaplane area where landing long and dropping down into the seaplane area would, no doubt, be equally as unthinkable as landing short. Buzz, like all the squadron pilots, knew the field well and knew what he had to do: not land short, not land long, and above all, clear the cliff to reach the runway area — in other words, make a near perfect precision approach without power. He would not be able to adjust or add power if he was going to be short, and he would not be able to add power and go around for another approach. His first and only approach had better be pretty damn good!

With all of this in mind, Buzz, with his dead engine and powerless but windmilling propeller, started his descending left turn towards the runway. What did not seem terribly complicated at first took on an additional twist when, not too long after beginning his turn, he realized that he was too high and would have to kill some altitude. He had essentially two choices: he could zig and zag (make some "S" turns) to lose altitude or he could make a three hundred sixty degree turn. A third choice, to "slip" in order to lose considerable altitude did not seem advisable in a military aircraft, so really was not thought worth considering. The "S" turns required space, and he could not be sure he would be in a spot at the end of the turns where he could get to the runway. Buzz opted for what he thought was his only real choice, a three hundred sixty degree turn, realizing that he could lose a lot of altitude in the turn and still not reach the runway. But he had no choice and would have to make the best of it. While in the turn, he wondered how things would be when he had completed the three-sixty, but he was pleased to find that he was in a good position altitude-wise and was fairly well in line with the runway he had to reach. As the seconds ticked by and he glided closer, his situation seemed less sure because he was losing altitude a bit faster than he expected. He was still sure he could clear the cliff, even if he had to touch down on the short grass strip before the edge of the runway. But a couple of seconds later, he knew for sure that he would either just clear the cliff — or that he would be just a few feet short and crash into the side of the cliff, the latter a crash that would surely kill all three men aboard. What to do? With no time to waste and calculating the dire consequences of crashing into the cliff and killing all aboard, Buzz made a sharp, steep, and, under the "dead stick" or failed engine condition, dangerous left turn and was in the water at the southwest edge of the field right next to the cliff where it starts to turn north along the west side of the field. Under the circumstances, they landed hard; but they all survived, although the crewman in the belly was injured and received a lacerated leg from the rocks which tore through the underside of the Avenger.

Placed in a situation not of his own doing and trying to cope with a field that offered no margin for error, who could find fault with his efforts and decision? He had made a valiant effort to save both the aircraft and the crew, but he faced odds no better than the flip of a coin, odds much worse than Russian Roulette. Who could argue that he did other than what was best under the circumstances? No one did!

Part Eight: More Surprises Bring More Challenges

More Surprises Bring More Challenges

In military aviation, as the reader has no doubt appreciated, incidents occur unexpectedly, some becoming quite serious, others more trivial. It is how the pilots react to those incidents that really makes the difference between a satisfactory resolution and a mess of serious proportions. Sometimes, the incident is caused by poor pilot judgment; and sometimes by faulty maintenance. And sometimes, incidents are caused by poor reaction to a problem — perceived or real — by a pilot who misdiagnoses the problem. And sometimes the weather, including icing imposes serious challenges, challenges most often, but not always, satisfactorily met largely through pilots' skills and experience.

But weather challenges, however demanding, constitute but one area of the many types of challenges that military pilots face. Often, just the nature of the beast, while tame and manageable in most conditions and circumstances, can, somewhat unexpectedly, become ugly and mean — and at the just the wrong time. Mentioned earlier was the relatively slow response of turbo jet engines to the sudden application of a fully advanced throttle. To avoid engine failure caused by too much fuel, a "rich out," the fuel fed by the throttle had to be controlled through a fuel metering device; this requirement added precious seconds to obtaining full power needed in certain situations.

Reciprocating engines also harbor a beast innate in their design — torque. While somewhat negligible in turbo jet engines, it is always present in reciprocating engines and is well-compensated for in design features of the wings and fuselage. But again, it is in the sudden application of full power where the beast asserts itself, sometimes with tragic results. Simply stated, it is the basic law of physics which asserts that, for every action, there is an opposite and equal reaction. In reciprocating engines with their big propellers, the clockwise spinning of the propeller blades creates a tendency for the entire aircraft to "react" by spinning or rotating counterclockwise.

As presented in the descriptions of incidents so far, it is easy to note that not all tight or dangerous incidents are caused by questionable pilot decisions, even if in hindsight many pilot decisions look questionable, perhaps even stupid. But, as in Dick Allen's case, many are caused by mechanical problems. Naval operational aircraft are very reliable, very rugged, and very durable — provided they receive the timely and proper maintenance specified. They can fly for thousands and thousands of hours without incident. They are also very complex vehicles with many systems and moving parts; but, from time to time, mechanical problems do occur. Anyone who owns an automobile, household labor-saving devices, or electronic equipment including computers knows that he has had his share of problems with the equipment.

Mechanical problems manifest themselves in several ways. New aircraft types, and later models of the same aircraft frequently have problems which only constant squadron operational use can reveal. These are routinely taken care of through the Navy's system of changes at either the squadron level, or if more involved, through overhaul procedures. Other mechanical problems come as the result of wear and tear on the aircraft as it ages in continual use, problems which, despite the Navy's very fine inspection and maintenance procedures, do occur. While it would be an oversimplification to state that such problems are not usually serious, serious problems do occur, even if rare; but the consequences are often not life-threatening.

Excluding structural damage and obvious fires, which are the most serious problems the military pilot must face and which are most likely to be caused by combat conditions, the manner in which a pilot reacts to a problem varies, with some pilots being more calm and rational than others. Also, some occurrences are so rare or even unique that there really is no preparation for dealing with them. A good example of well-intentioned but unsettling things that can happen through carelessness had to do with the two following incidents having to do with the air conditioner and the tail pipe temperature gauge:

The sky had just turned completely dark on a moonless night, as Lieutenant Junior Grade Harry Martin started his takeoff run in his swept-wing F9F-6 Cougar. Airborne, but before he reached the end of the runway, and at a speed which precluded aborting the takeoff, without crashing through the cyclone fence at the airfield's border, a bright yellow light on one of his instrument gauges starting flashing at him. Now, just a hundred or so feet above the trees off the end of the runway, Martin was sure he was in serious difficulty. The pulsating light was impossible to miss and was indicative of trouble. "Oh no! What a spot to be in, and at this altitude," thought Harry. He quickly saw that the light came from the Tail Pipe Temperature Gauge, indicating overheating and, quite possibly, fire in the jet engine. Unquestionably a serious problem when you have only one engine! The possibility of engine failure or fire made him briefly think, " This is how it happens. You lose your engine and are too low to bail out, so your time has come." Either climbing to a safer bailout altitude or getting back down on the runway fast were his choices. But the aircraft was still functioning, and there were no other serious indications, so he chose to stay with the aircraft no matter what. He throttled back a bit and turned downwind for an immediate approach to landing, dumping the fuel in his wing tanks on the way to get his aircraft down to acceptable landing weight. With the throttle pulled back as he started to letdown and make his approach turn toward the runway, LTJG Martin noticed that the flashing light went out. "What to do?" thought Harry. He had already dumped the fuel that could be dumped. And how could he be sure that the light would not flash again once he added power to climb to the operating altitude? "No. There was some sort of problem, so it was best to land safely and as soon as possible and check it out."

He continued his approach, landing safely; and while still taxiing to the flight line, the light remained out. Even while still taxiing, he seemed to sense what the problem was; and it did not make him feel good, even maybe a bit foolish. After parking and cutting his engine, he took a good look at the Tail Pipe Temperature Gauge, a look that the pulsating yellow light had precluded him from making in the air. Normally, the fixed needle is set and fixed on 816 degrees Fahrenheit as the limit above which conditions in the engine and exhaust system are dangerous and the warning light begins to flash to indicate a problem — certainly a situation in which landing as soon as possible would be a good idea, and a situation which would have justified Harry's landing immediately. In this situation, for some unexplained reason, THE FIXED NEEDLE WAS INCORRECTLY SET AT 618 DEGREES, which is the normal takeoff temperature reading and is the point the moveable temperature needle reaches on every takeoff. In the existing conditions of Harry's night takeoff and the pulsating yellow light, Harry could not, or did not, immediately see that the takeoff tail pipe temperature limit needle was set incorrectly at only 618 degrees. Harry did feel a little embarrassed, but consoled himself by knowing that he had never heard of an incorrect tail pipe temperature setting — nor for that matter had anyone else, nor was it ever suggested that tail pipe temperatures gauges had ever been set at the wrong point. Harry did rationalize that, after all, he did not panic but acted prudently in getting the aircraft back on the ground quickly and safely to report and correct a problem not of his making. How the fixed limit needle happened to be set incorrectly, LTJG Martin never learned.

Another example of a pilot's reaction to an unusual circumstance is worth the telling. Lieutenant Wallace Parker, flying the fourth aircraft in his division, had barely taken off, when he thought he saw a little smoke coming from the console panel on the left side of his cockpit in his F9F-6 Cougar. Concentrating on keeping his eye on the three planes of his division in the air ahead of him so that he could rendezvous with them, at best he could only steal a glance at the console to see if his impression of smoke was imaginary or real. As he slowly moved closer to the other aircraft and was not in danger of losing sight of them, he could afford to take a somewhat longer look around the cockpit. Much to his surprise, he found a gray smoke seemingly coming from everywhere inside the cockpit. Though he was an experienced Navy pilot with many hours of flight time, he had only recently transitioned into this jet. To him, smoke meant a problem. Though there was no odor and the smoke did not get in his eyes, he felt that, despite his strongly felt compulsion to complete the exercise, he should advise the division leader and abort his flight. Parker radioed, "Cougar One, this is Cougar Four. I have smoke in the cockpit and am returning to base." It seemed such a routine procedure to advise the flight leader and such a good idea to return to base under the circumstances that he was not really prepared for the flight leader's response which seemed condescending, even sarcastic. "Turn off your air conditioner!" Parker did switch off his air conditioner, and, sure enough, the smoke or really mist, immediately started to disappear. "How could I be so stupid," thought Parker, resuming his rendezvous. "Such a simple thing to know, and I didn't know it."

In spite of feeling embarrassed and a bit naive, LT Wallace Parker promptly put that feeling aside and joined the formation to proceed with the mission. Later, Parker was to wish that his air conditioning problem was as simple as turning off the switch, as the flight leader had advised, and that the problem had really been resolved. With the mission completed, the flight leader advised the group that they were to break up and return to base independently. During the flight, Lt Parker had noticed that some ice had gradually begun to form on the inside of his windshield and canopy; and after the division broke up, the ice formed more quickly and heavily. Aside from very much restricted visibility when you are travelling at 600 miles per hour, a serious hazard in itself, there is no other real danger provided you can see well enough to avoid a collision. Ice will melt once you descend into warmer, dryer air. It just takes a little while, longer if, as in this case, the ice buildup has been thick. But to his satisfaction, Parker was able to complete the mission despite the temporary diversion.

Part of LT Wallace Parker's embarrassment was due to his not knowing beforehand of the conditions which sometimes affected this particular air conditioner and what to do about them. Certainly he had not read about it in the pilot's handbook, and certainly nothing was said by the squadron's aircraft type officer, who was supposed to be the squadron expert on Cougars. So, was it something insignificant, nothing to worry about? Of course in this case, it presented no real danger — provided you know about it and realize that it was not smoke but mist. But unfortunately, not everyone knew about it; and there was a report that one pilot making a turn into final approach for landing, a position in which the pilot was low and not too much above the stalling speed, saw smoke in the cockpit, reported it to the control tower, and then, in the concern of the moment, stalled and spun into the ground with the usual results. Reports from investigators determined that it must have been mist and not smoke from a fire. Just how the pilot reacted to cause the accident is not known, but it is likely that, at the least, he was distracted at a particularly inopportune moment. How frequently the air conditioner behaved in such a manner or what atmospheric conditions might have contributed to the creation of the mist most Cougar pilots seemed to know. But many Cougar pilots knew that they had never experienced it even once.

Further evidence of the surprises that can accompany all pilots, not just Navy pilots, are numerous. The important thing is that, in almost all unexpected situations, patience, experience, and a calm approach to their solution brings about the wished-for resolution. One thing is sure: Pilots will continue to encounter surprises, even mild ones like these.

A Fine Mess We've Gotten Into!

This oft-used line of Oliver Hardy's when speaking of a predicament he blames on his sidekick Stanley Laural, " A fine mess you've gotten us into!," may not exactly fit the circumstances relayed in the situation described in this section. If not, perhaps one more apt could be substituted.

Throughout these episodes, weather has reared its head again and again. But it is only because weather is such a factor in aviation. Even if passengers in a commercial airliner are not always aware of it, their flight may be contending with challenging weather conditions on takeoff, in the climb, en route, or at the destination airport — or perhaps during all of these phases of flight. The state of electronic navigational aids, radar surveillance, landing approach systems, etcetera, make flights in bad weather somewhat routine for the most part. But extreme weather conditions are still very limiting. Everyone has seen how the wind can uproot trees or snap them off at mid-trunk; they have also seen how heavy, wet snow or the clear ice from freezing rain has broken huge branches off trees or snapped power lines. Everyone knows about tornadoes and their fury. Somewhat less known is the severe turbulence of thunderstorms and a phenomenon, still under much study, called clear-air turbulence or simply CAT. These are conditions to be avoided, if possible, in the air; but they are much more serious during takeoffs and landings. Of serious concern during the landing approach are the violent "sheer forces" of thunderstorms, in which violent updraft and downdraft forces act on the aircraft alternately or possibly at the same time. Most of the severe forces are very infrequent; the likelihood of encountering sheer forces on landing are slim, and current electronic equipment is often able to detect them or anticipate them and advise the pilots. And given reasonable attentiveness, pilots can operate their aircraft safely in most conditions, flying around or waiting until the weather condition passes, since most severe weather moves on in a relatively short time.

Military pilots likewise must be able to operate in all kinds of weather and at all hours of the day. In addition to all of the electronic navigational equipment available to the airlines, Navy pilots have access to radar-guided approaches to landings in very poor weather at military airfields. This, of course, is the Ground Controlled Approach radar described earlier — GCA. Part I described how Navy aircraft descended through dense clouds through as much as thirty-five or forty thousand feet of altitude. It was clear that a serious concern about those letdowns was the amount of fuel left for landing. Since military fighter and attack aircraft do not have the large reserves of fuel available to airliners, their pilots have to be sure that weather conditions below remain at or above the minimum conditions for landing. If they are to fly to some alternate airfield, it is much, much better to proceed at altitude where fuel consumption is much lower. In mid-ocean, there may be no alternate airfield within range; at sea, Navy pilots often make excellent approaches and landings when weather conditions are extremely undesirable, even below established minimum standards — simply because there are no options.

With all of the fine navigational systems available to pilots today including omni-directional-distance measuring systems (OMNI-DME), a similar military system called Tactical Air Navigation (TACAN), and satellite navigational systems providing ground position information, it is easy to think that such systems were always available. As wonderful as these systems are for navigation, some are even more useful to pilots flying who have to make a letdown to a runway in minimal conditions on instruments. Besides being aids to navigation, they are aids as well to confidence and a feeling of security.

But things were not always so. Not too many years ago, were the relatively primitive days of "on the beam" flying associated with often unreliable low frequency radio ranges, where A (dot .) and N (dash —) Morse Code signals from adjacent quadrants overlapped so that together they formed a continuous "on the beam" sound indicating headings to or from the transmitting station. It was this system, with its idiosyncracies which was reported to be a factor in a B-25 Mitchell Bomber's crashing into the Empire State Building right after World War II. Another navigational instrument was the Radio Direction Finder (RDF), an antenna loop which could be rotated to locate a radio source, the signal from the source being strongest when the loop was at right angle to the radio source; but this device could not tell the pilot whether the signal was fore or aft, an additional test being necessary to determine that. Quite an improvement was the Automatic Direction Finder (ADF) which would actually point to the signal source, eliminating the searching and the ambiguity of Radio Detection Finders's not providing information as to whether the signal source lay in one direction or the other. Today, systems based on the ADF idea take many forms, and often include Distance Measuring Equipment to indicate how far it is to the signal's source. These systems, of course, are invaluable; but however valuable these systems are, they become useless if the transmitting source is inoperable or if the aircraft's equipment is malfunctioning — or, if in case of wartime necessity, the signal transmissions have to be shut down. With these difficulties, pilots must revert to more distant times and older, more primitive, but still reliable methods of determining their location or position such as compasses — or celestial navigation in the case of big aircraft, a method hardly, if ever, used today, although some navigators routinely practice the procedure should the actual need arise.

There are many tales of pilots flying in inhospitable environments over long distances with less than satisfactory means of navigation. Visibility, whether during a black night, clouds, fog, snow, rain, icing, or conditions known as "whiteouts," could necessitate flying on instruments. But additionally, what does one do when he can't climb over a high mountain range or letdown in mountainous regions with unreliable equipment? Then he must do everything he can from descending through clear spots to letting down over the water, hoping to reach his destination visually. Neither of these adaptations is without risk. In addition to those problems, the pilot may have to deal with ice that forms on the aircraft, particularly freezing rain which the pilot might not notice immediately and which can add significant weight to the aircraft.

Today, however, greatly improved navigation systems are a Godsend, as are the many excellent airports with reliable letdown equipment around the world; and the power and speed of turbojet aircraft enable pilots to reach extremely high altitudes and climb through levels of freezing rain quickly to minimize accumulation. So the kinds of experiences like those in Ernest K. Gann's book *Fate Is the Hunter* are very uncommon today. But they still can occur and one of the incidents which Gann describes still has meaning. Pilots with whom Gann was associated were flying transport missions from Presque Ilse, Maine, to Goose Bay, Labrador, proceeding on to some of the quickly established bases like Bluie West One and Bluie West Eight in southern Greenland, and then on to Reykjavik, Iceland, during World War II. After leaving Bluie West One, pilots had to fly across Greenland's vast Ice Cap en route to Reykjavik. Gann describes flying across the ice cap both in the clouds and in the clear.

"For here was enchantment such as few men had yet beheld. The ice cap surged upward from the overcast in a great rumpling tidal wave of snow. The protruding mountain peaks we had seen from afar now appeared much higher than had been alleged. A few most certainly bettered ten thousand feet. We had been warned to approach the ice cap with care, for in certain combinations of light it merged so gradually with the sky that there was no visible horizon. And since no one professed to know its exact height, maximum possible altitude was also recommended. On this morning, there did appear to be a faint division of light marking a horizon, but it was ephemeral and untrustworthy. So as (the pilot) Johnson turned northeastward we resolved to fly on instruments, rejecting the invitation to depend on our eyes. Since we were quite in the clear, the sensation was deeply disturbing, like flying toward a constantly retreating wall.

"And as we continued, assimilating little by little the dominant hostility of this land, we knew the sensation of being dwarfed to nearly zero proportions; we became far more diminutive than even a thunderstorm could render us. We were less than gnats in the sky and were accordingly cleaned of ego."

Gann's descriptions of the vastness of the area, the hostility of the cold, the troubling yet enchanting environment, and the pilots' thoughts of their own insignificance, help to set the scene in this description which Gann relates: "One of the strangest 'crashes' in aviation history occurred on the ice cap when the pilots of a PBY (Catalina) elected to fly across, relying entirely on their senses. The first indication that anything was awry came through a lack of sound and airspeed. With engines turning at full cruising speed, they literally flew onto the ice cap, touching it so lightly at first that their speed was gradually reduced until they were at a standstill. Fortunately, they had slid upon a smooth area, and when they finally realized their predicament, they stepped out laughing into the snow. Their laughter was short-lived. They were very much alone with a useless flying machine and were a long time being rescued." It really was a fine mess they had gotten into!

Nearly an unbelievable occurrence, it does add one more strange incident to the large store of events which make up the sum total of flying experience. Needless to say, many strange happenings peculiar to aviation end not so benevolently. But again, it is not the purpose or the theme here to dwell on the tragic, but rather those occurrences in which the pilots persevered to resolve the challenges of life-threatening situations.

Bad weather does not only mean flying through dense clouds from practically the ground up to 35,000 or 40,000 feet and making instrument letdowns to satisfactory landings in minimum or near minimum conditions of ceiling and visibility. Snow and ice buildups on the aircraft, and snow and ice buildups on the runway are part of the weather package which pilots must contend with. Powdery snow on the runway presents less of a problem for landing than for takeoff, while an ice-covered runway presents more of a problem for landing than for takeoff, as in the landing on Adak's ice-covered runway, related below. And a runway that has equal patches of ice and dry ground can present a problem when the pilot brakes both wheels, only to have one wheel brake on dry ground while the other slips on the ice, causing a tendency for the aircraft to veer to the braking side. Anyone who has driven an automobile on an icy road knows the hazards involved in attempting to brake the car to a stop. Many times, if braking is possible at all, the car will tend to swerve, requiring the driver to ease up on the brakes and turn the steering wheel to counteract the swerve. If the car hits a dry patch of road during the swerve while the wheels are not headed in the direction of the motion over the ice, it can roll over. For aircraft and their high speeds at touch down on the runway, snow and ice present bigger challenges. A military jet fighter without a drag chute needs brakes to stop. The jet engine, even at idling RPM's (power), will keep the aircraft rolling endlessly down the runway after touch down, so conditions in which brakes can be applied effectively are essential. Wet, slick, short runways can present similar braking problems. Bigger multi-engine aircraft have reverse pitch propellers or, in the case of big jet aircraft, reverse thrust devices they can employ to help slow down the aircraft after touching down on the runway. Even so, there have been numerous cases where commercial jet aircraft have had trouble stopping on icy runways, an example being a relatively recent instance in which a commercial jet skidded off the end of the runway and into a river. Add a brisk crosswind to an icy runway, and you have an additional concern. Fortunately, the clearance of snow and ice from runways receives a very high priority, greatly reducing the conditions which can cause problems. But sometimes pilots do not have the luxury of having a dry or an ice-free runway.

For years, a story, repeated many times but unconfirmed by anything the author has seen in writing or knows from personal contact, made the rounds about a P2V Neptune landing incident at the Naval Air Station in Adak, Alaska. The runway in use was very icy and not very long at about 5,000 feet. The wind, though not strong, was a sixty-five degree crosswind from the right, and from that direction did not have to be very strong to cause problems. The pilot's approach to the runway was good and the aircraft touched down in the first five hundred feet of the runway. At the right moment after touching down, the pilot reached up behind the throttle knobs to the reversing levers to reverse the pitch on the propellers and increase the power which would then push the air away and ahead of the plane to slow or brake the aircraft's speed somewhat. Since the braking action of reverse pitch propellers or reverse thrust jets is most effective soon after touching down and getting the nose wheel on the ground, and while the aircraft is still at a relatively high speed, that is when it is employed; very shortly thereafter it is discontinued and brakes are engaged as necessary.

On this landing, either as the pilot was coming out of reverse pitch or shortly thereafter, the nose of the aircraft started to head to the right and into the wind. At the same time, the aircraft was being pushed slightly to the left side of the runway by the right crosswind. Since the surface was seemingly all ice, more than that which was reported, braking was useless; and the use of left rudder to correct the nose heading would only aid the crosswind in pushing the aircraft further to the left side of the runway. The right throttle could be used to straighten the plane down the runway; but that would increase its speed, and the runway length made that unwise. As the plane was still heading more and more into the wind, it was still moving or sliding more or less straight down the runway. Although unlikely in these apparent conditions, should the aircraft hit a big dry patch while sliding this way, severe problems could arise. At what point the pilot decided to experiment with the left throttle is not known, but apparently he felt that his best chance of braking the aircraft safely to a stop lay in completing a full one hundred eighty degree turn so that the Neptune would be facing the direction from which it came while it was still sliding backward in the landing direction. The pilot carried out his plan successfully, aided of course by the presence of ice sufficient to let him complete the turn. But once headed in the direction from which he came, all he had to do was to advance both throttles evenly and with sufficient power to bring the aircraft to a complete stop. The accommodations which Navy pilots are forced to make in circumstances forced upon them are truly interesting and prove that there is very, very often an escape from "The fine mess they've gotten into!"

Things You Just Have To Respect

Other than mechanical and aeronautical engineers and mechanics using torque wrenches, most people have only a vague idea of what "torque" is. Torque can be very useful. Torque can be dangerous. In aviation, the use of torque can be very challenging. In aviation torque is the means of motion, the means of delivering the power in engines. Properly applied, it can do wonderful things. Improperly applied, it can create disaster.

Defined simply, torque is the turning power or the turning power applied to a shaft. Strictly speaking, shaft power is not only created by a fuel-burning engine. A crank handle and shaft which raises a bucket of water from a well delivers shaft power; so also does an electric motor. However, the kind of power referred to here is that which is created by powerful, fuel-burning, aircraft engines — and particularly reciprocating engines, those whose back and forth motion of piston rods is turned into the rotary motion of a shaft which drives a propeller.

If a propeller were attached to an aircraft body or fuselage that had no wings or tail, the propeller would turn in one direction and the fuselage would rotate in the opposite. Thus, while the propeller is turning clockwise, an equal and opposite force turns the fuselage counterclockwise. It is the presence of the wings and the design of the tail, particularly the vertical section of the tail called the vertical stabilizer, which keep the aircraft from turning in the opposite direction of the propeller. Torque considerations and design are equally important for helicopters, where without the tail rotor, the large lifting blades would turn in one direction and the helicopter body would rotate uncontrollably in the opposite direction.

In propeller-driven aircraft, the vertical stabilizer, which appears to be lined up evenly with the center line of the aircraft, is actually offset a little bit to one side of the center line. That offset is designed to compensate for the effect of the powerful blast of air created by the spinning propeller. With a clockwise-spinning propeller, as most are, a constant blast of spinning air or swirl affects the left side of the fuselage and tail, or vertical stabilizer, with a strong portion of the swirl striking the tail after it has gone under the right wing and spins up against the left side of the tail, both swirls tending to push the aircraft into a left turn and roll. The reader will quickly deduce that an aircraft on the ground and not moving or moving slowly during the first part of the takeoff run, will not have much airstream to counter the propeller's blast. So, a problem can develop here. Problems can also occur when an aircraft is in slow flight as in landing at a field or in an approach to landing on an aircraft carrier, when the pilot has to make a sudden application of power. Torque then is at a maximum, likewise the blast of swirling air from the propeller — all at a time of reduced speed and minimum airstream strength.

With some few exceptions, World War II Navy fighters, dive bombers, and torpedo bombers were powerful single-engine, low wing aircraft with nose-mounted radial engines, with big propellers. This configuration resulted in an aircraft design with a tail wheel rather than a nose wheel. Aircraft design till that time mainly incorporated a tail wheel. Even the Navy's workhorse, the R4D Skytrain (Army's C-47 Dakota, Civilian DC-3) and the legendary B-17 Flying Fortress had tail wheels. In the present age of jets, it's hard to believe that the Navy's only big single-engine, big propeller aircraft of extensive use which had a nosewheel rather than a tail wheel was the T-28 Trojan. The F7F Tigercat had a nosewheel but was a twin-engine aircraft; and the Ryan FR-1 had a tail wheel, but it had both a piston engine and a jet engine. The significance of having a tail wheel was that, when the aircraft was on the ground at slow speed, the tail wheel of necessity had to be on the ground to prevent undesired turning due to torque, spinning air blast, or crosswind.

Although at one time a "three point" landing (three wheels touching the ground at the same time), was considered and was in some instances a good technique, particularly for carrier landings, the "three point" takeoff, while possible, had a disadvantage. The more common practice was to raise the tail wheel, and hence the tail, as soon as possible and roll along on the two main wheels. This method was preferred because the wing would be parallel to the ground and presented less drag or resistance to building up speed for takeoff. It is not surprising then to note that many pilots, in their attempt to get the tail up and build up speed, raised the tail wheel off the ground too soon, with the result that the aircraft veered sharply to the left, in some cases even going off the runway before becoming airborne. While this inclination was not entirely the result of torque, torque was a factor in creating the swirl of air from the propeller which was forced down and under the right wing only to come up and strike against the left side of the vertical stabilizer area, pushing the tail to the right and forcing the nose to the left. Some aircraft were more prone to this type of problem than others: the Navy's SNB/JRB Beechcraft (Army and later Airforce C-45), and the Navy's Howard, both utility aircraft. This problem often occurred more frequently to pilots who were qualified in many different aircraft and who had to remind themselves of this strong tendency in certain aircraft. Although the potential for a serious accident and injury was present in such situations, the author knows of no serious accidents or fatalities in such incidents.

Far more dramatic and serious are the incidents involving the sudden application of full power when the aircraft is slow and is also low as in a carrier landing approach or in the practice carrier landing practice mode called FCLP, where the pilot is making practice approaches using part of a runway and having a landing signal officer (LSO) guide and monitor his approaches. A couple of happenings illustrate more clearly than lengthy explanations.

Lt. Mac Hartzel was a good, competent pilot. He was positive and upbeat; and, as a senior lieutenant, he set a very good example for the younger, less experienced pilots. Respected for his knowledge and ability, he was equally respected for his cheerfulness, his encouragement of newly arrived pilots, and for his general ability to create a team atmosphere and a spirit of pride.

In getting ready to go aboard the aircraft carrier, the *USS Coral Sea*, Mac led a two division flight of eight aircraft to an outlying field of Naval Air Station Norfolk, Virginia for field carrier landing practice, or FCLP, in AD-1 Skyraiders. The landing signal officer, or LSO, and his assistant, who would take notes about each approach as dictated by the LSO, were in place at the downwind end of the runway, or the simulated carrier deck. Mac led his two divisions upwind along the right side of the runway, the aircraft breaking off individually just past the end of the runway — exactly as if they were at sea and going to fly abroad.

The practice seemed to be going fairly well, each pilot making approaches varying but little from one approach to the next and being guided by the LSO's flag signals. The LSO's flags were always at arm's length except when he signaled for a cut sign or a wave-off, the cut sign, or cleared to land sign, being indicated by his taking his right arm and flag sharply across his chest and the wave-off being waving both flags sharply, and repeatedly, across his chest, an indication that the pilot must not land but go around for another approach. A "Roger" signal, when both arms were level and parallel to the deck (just like the letter R when signaling with semaphore code flags), indicated everything was just right. Raising the arms somewhat indicated a high position and dropping the arms slightly indicated a low position. Other signals indicated either a too fast or a too slow speed. This procedure represented teamwork: the pilot would make his approach knowing what he had to do, and the LSO would assist, using his judgment to determine whether the pilot had his aircraft at the right place, the right height, and the right speed to receive a cut signal. Right or wrong, the LSO's judgment had to be taken as law. If he indicated a cut, the pilot reduced power abruptly and landed; in these practice cases at the field, the pilot would land, add power, and takeoff for another approach. If the LSO indicated a wave-off, the pilot was required to go around for another approach.

As Mac was making his fourth approach after three good approaches, he saw nothing to indicate that this approach was the least bit different from his previous, quite satisfactory approaches. But just as he reached the spot where he expected a cut signal, he received instead a frantic wave-off signal from the LSO. In anticipating another cut signal, it's possible Mac throttled back just a bit. No matter. Disciplined to obey the LSO, Mac applied full throttle to his three thousand horsepower engine; the aircraft turned sharply to the left, actually rolled to the left to a point where the wings were nearly at a right angle to the runway. With very little vertical lift in this position, the plane dropped and the left wing struck the ground. Aided by the engine's power now being fully effective in increasing the plane's speed and by Mac's corrective action, the aircraft bounced into the air with a damaged, partially bent up wing. Miraculously, no other part of the aircraft was damaged or even scratched: not the flaps and not the landing gear, and Mac was able to fly a damaged, but relatively safe aircraft, back to Navy Norfolk.

There was no need to fault either Mac or the LSO exclusively. Such incidents usually occur partly as a result of both the pilot and the LSO and must be considered as a learning experience for both. The system, the best known at the time, was not perfect; and neither man was careless. Most pilots knew by feel and perception whether their approach was good. On the other hand, it was not unusual for pilots to anticipate one signal and get another. LSO's, being human, make mistakes also. They have to judge by the plane's attitude (proper nose up position) to know whether the speed is about right. And since the weight of different versions of the same model can vary, the correct attitude can vary; and with an incorrect attitude, the speed can be too fast or too slow, a critical factor in carrier landings. Sometimes, an LSO might give a cut signal to a pilot who is slightly high, thinking perhaps he is not too high. The result can be that the pilot feels he must "dive for the deck" to land within the allowable space; such can be the cause of serious problems. But the bottom line in this incident was that, whoever was at fault, perhaps both, Mac lived to fly another day and move on into the jet aircraft era and become a good skipper of a jet attack squadron.

In a similar, but more serious and spectacular incident, young Ensign Neil Butts was returning to the *USS Midway* aircraft carrier after a routine flight at sea. Attached to a fighter squadron flying Gruman-built F8F Bearcats, Butts was, as was everyone else in the squadron, fairly new in the aircraft. The Bearcat was a very impressive hot-looking airplane that seemed to be all engine and practically no vertical tail section. It had wing tips that were designed to break off during certain high G (gravity) maneuvers. On a bright sunny day, Ensign Butts was making his approach to the carrier and was very close to the ship. He was in that critical spot where a cut signal or a wave-off would be expected. Perhaps he, like Mac, was expecting a cut signal and had unconsciously eased off some throttle. Perhaps the LSO was fearful that, if Butts continued as he was, he would strike the ramp — that downward-curved, aft end of the flight deck — with catastrophic results. All is conjecture here because the incident, as usual, was not widely discussed with pilots in other squadrons of the air group. At any rate, Butts received a wave-off at the last possible second and applied full throttle to his Pratt and Whitney R28-34E with between 2150 to 2500 horse power engine. Again, the result was a turn and roll to the left; and when the wings were perpendicular to the deck and there was no vertical lift, the plane slid down and the left wing tip hit the deck. In Butts case, he barely bounced a few feet above the deck, his roll continuing until he was on his back, or upside down and vertically heading straight nose down. With a big splash, he crashed nose down in a near perfect dive into the ocean off the port side of the carrier. No one dared hope he could survive. Navy personnel in the catwalk on the port side of the carrier sort of expected some kind of explosion or aircraft disintegration from the tremendous force of the crash. Everyone was fearful of what he might see. Everything happened in two seconds; it seemed incredulous.

In what had to be one of the most unbelievable crashes to witness, Butts and his aircraft, after crashing nose-first into the ocean, popped up like a stick thrown into the water pointed end-first. Like the stick, the F8F popped back up so that the tail and even the cockpit were above the water; and Butts could be seen getting out of his shoulder and seat belts, inflating his Mae West, and scrambling over the side of the cockpit and into the ocean clear of the airplane, which sank immediately. One cringed to think what might have been the result if Butts had been knocked unconscious or even momentarily stunned. Sure, Butts had gotten out of the aircraft and into the water. But what was his condition? Surely, he had been injured, perhaps seriously. Perhaps with broken bones. Perhaps with severe internal injuries. Perhaps with severe head injuries — after all, at that time helmets were not worn. Without question, he had been injured, but he was alive! Butts was picked up almost immediately by a helicopter, flown back on the carrier and taken on stretcher to the sick bay, where doctors determined that he had been badly bruised, but that he had no broken bones or head injuries. And what internal wrenching there was, was not serious. After being tended to in the ship's sick bay and kept for observation for a day or so, Butts could be seen on the flight deck, walking about with the aid of crutches. His survival was testimony to his basic good condition and, perhaps above all, to his youth. His youth would also help him heal quickly; but even so, it would take awhile before he had recovered sufficiently to receive medical clearance to put him back in a flight status.

It is noteworthy that Jane's *All The World's Aircraft* has this note about the F8F Bearcat: "Last of the single seat, single engine, fighters to originate from the F4F Wildcat; one of the most successful piston engine aircraft to serve in the Navy." More pertinent in view of this incident is that it (the F8F Bearcat) had a "small tail but the fin and rudder height were increased to improve directional control."

Some pilots will swear that accidents like this are the fault of the LSO's. Some LSO's will say any accident like these is the pilot's fault, not theirs. Others will say that each is partly responsible. And sometimes the LSO is faced with a situation, in which, however the situation developed to a certain point, any decision can be a bad decision. Regardless, in these incidents, both pilots survived. It might have been that the more experienced Mac eased the throttle to control the torque, and the younger, less experienced ensign did not or could not — but who is to know and who can judge in hindsight? What is known is that Mac Hartzel was up and flying the next day and continued in a long flying career. Though not known by the author, it would not have been unusual to learn that Ensign Butts, after rest and recuperation, had done the same.

With the transition to the faster jet aircraft, adjustments became necessary. Experiments with new devices to make the system better were tried. A lighting system to replace the LSO's flags was tried; but this was operated by the LSO, and really was not much of an improvement. An angle of attack indicator, affixed to the aircraft served to tell the pilot whether his approach attitude, and hence largely his speed, was correct. A big improvement was the mirror landing system, which, through the use of a big mirror, lights, and electronics, enables the pilot to know when he is on the precise approach required for a safe landing. Later, and still in use is The Optical Landing System and The Improved Fresnel Lens Optical Landing System.

No system, however, can be perfect, particularly in rough seas when the carrier is rolling or pitching, or even rolling and pitching, or when the weather conditions are bad, and at night — and when, as occasionally happens, when all of these conditions prevail simultaneously.

With Naval Aviation and in attempting to land aboard aircraft carriers, the pilot is essentially landing on top of a plateau, some sixty to seventy feet or so above the water. To do this, the pilot must have his aircraft slowed down so that, when given the "cut signal" by the landing signal officer (LSO), his pulling the throttle off will let him touch down in the first section of the carrier deck. However, he cannot touch down too early for fear of not clearing the ramp, that curved-down first part of the deck or "ramp." One need not have to imagine too much to realize what the consequences of hitting or "striking" the ramp can be. If most of the aircraft is clear of the ramp, with only wheels or tail surfaces striking the ramp, the pilot might survive an incident in which the rest of the aircraft breaks up. So it is the demanding nature of the approach, including precise timing, which presents a situation, slight variation from which, can cause serious problems. That so many thousands of carrier landings are effected satisfactorily, is an indication that the system almost always works.

There remains a tendency, however, for a number of pilots approaching the carrier to think or anticipate that they are about to get a "cut" signal before the LSO is quite ready to give the "cut' signal. The pilot sees his position as such that, if he cuts his throttle, he will be in the desired place on touch down and often eases off the throttle. One of the LSO's biggest concerns is with having the approaching aircraft clear the ramp safely. Observing a dangerous situation's developing and not wishing to risk the serious consequences of a "ramp strike," the LSO is in a difficult position: will the pilot clear the ramp when he gives him the "cut" signal? Perhaps he will but not wishing to take the chance, he will give the pilot a "wave-off" signal to go around for another approach. Already slow, and perhaps dangerously low, the sudden application of full power creates the worst elements of the resulting torque: not only is the aircraft starting to roll counterclockwise in resistance to the propeller's clockwise rotation, but also there is the swirling slipstream created by the propeller which curves under the right wing and comes up and against the tail's vertical stabilizer, exacerbating the roll and, in addition, forcing a turn to the left. The effect of all of this is that, at the same time the aircraft's wings provide reduced lift due to the slow speed, the wing's position at forty-five or more degrees to the horizontal provide only half the vertical lift of horizontal wings. There has to be a loss of altitude unless and until the pilot can reduce power a bit, level his wings, and slowly build up the airspeed necessary for level flight. The examples described here are indicative of the general problem, with the best solution being a complete understanding of the problem by pilots and LSO's.

Although jet aircraft have much less torque in those situations than propeller types, they do have engine acceleration problems and can end up in serious risk of "ramp strikes." Over the years, there have been many "ramp strikes," or even accidents in which the approaching aircraft struck the carrier below the ramp and hit the fantail of the ship. Some incidents are more serious, even disastrous; but always there is serious aircraft damage, and in most cases, there is bodily injury or worse. It is, perhaps, worth noting that one widely circulated report describes an incident in which a pilot making an approach to the carrier struck that part of the ship below the ramp and above the actual fantail — AND LIVED! The nose of his F2H-2 Banshee miraculously crashed into a narrow corridor; and in the process sheared off both wings, leaving him safe and secure in the corridor and in his cockpit. Given the nature of this incident, those described above, and many others over the years, is there any doubt that avoiding potentially dangerous situations in the first place is preeminent?

* * * *

Believe! Your Life May Depend Upon It

Mentioned earlier was Rocky's graveyard spiral incident. The essence of the "spiral" is that it originates when the pilot blacks out when blood drains from the pilot's head during a high-G turn. The exercise that requires such a tight turn is what is called "the dog fight." Fight tactics are adapted to the situations encountered. With changing tendencies, the typical dog fight may be less frequently used today. But anyone who has seen the movie Top Gun cannot forget the extreme maneuvers used by those pilots in their attempts to gain an advantage over the enemy. And today, fighter pilots are always seen wearing G-Suits or "anti-gravity" suits to keep from blacking out in tight turns. Surprising to many is the note that the Navy's flight demonstration team, the Blue Angels, does not use G-Suits, despite some very tight turns. In flight training, it serves a useful purpose to let fledgling pilots experience blackouts. However, there is an attendant problem that goes with blackout training: pilots who disregard the preflight training or who think they know better can get into life-threatening situations. Most pilots take their training very seriously; but occasionally, someone like Rocky Levine seems to ignore the training or become confused when he blacks out and has the sensation of losing altitude. Feeling that he is losing altitude but has not yet regained sight to observe his condition, he pulls back on the control stick, keeping the blood from returning to his head and keeping the aircraft in other than level flight. While not an instrument problem, it is similar in that the pilot has to have confidence that certain actions will work. The blacked out condition affects only, or nearly only, the pilot's vision and not his mental processes. So, the pilot only has to realize that easing up on the control stick makes the turn less tight thereby reducing the G-forces and letting the blood return to his head, restoring sight. Apparently, in so many blacked out situations there exists the sensation that the aircraft is losing altitude quickly, maybe even that the aircraft is in a steep dive. The mistaken remedy for the pilot so affected is to pull back on the control stick to recover from the dive. That remedy keeps his turn tight, his head blacked out and his aircraft losing altitude all the way to the ground. Hence, "The Grave Yard Spiral."

Even though the pilot in a grave yard spiral is blacked out, his status is not unlike the condition of a pilot experiencing spatial disorientation in clouds. Both seem to rely on physical sensations; both try to effect a change, albeit the wrong change, by ignoring the lesson just to take their hands off the flight controls and let the inherent design features of the aircraft tend to right itself. In both situations, the value of the instruments is minimized, so the pilots persist in their actions. For the blacked out pilot, getting his sight back so that he can see where he is and how to level his wings is important; and seeing his instruments is crucial if he finds himself in the clouds. Naval Aviators know that Naval aircraft are basically stable and that just releasing the control stick will bring about some improvement in the aircraft's flight attitude. But essential to recovery is the prompt realization of the condition the pilot is in and then the application of the procedures he has been taught. It is a strange and ironic thing that the pilot feels he has to take action to keep himself both in the grave yard spiral and the disoriented state, when doing nothing, that is releasing the controls will restore more level flight — provided the action is prompt and there is some altitude to effect recovery.

In all likelihood, the pilot in a grave yard spiral is not in a turn approaching ninety degrees. He can be losing altitude in most turns when the wings are not level; but the likelihood is that when the pilot's turn is sufficiently tight to keep blacked out, he does not have enough lift and may descend rapidly. In Rocky's case, he and his opponent had been briefed by the instructor who later accompanied the student-pilots, and would evaluate their skills. Rocky had been a fairly good student pilot and gave no indication that he did not understand what was required and the precautions to be taken. But on his very first attempt to get on his opponent's tail, something happened to Rocky. Was he too eager in his attempt? Had he not been paying attention during the preflight briefing? Whatever, Rocky continued in a tight turn and started to lose altitude. His actions did not immediately cause concern because everyone believed that Rocky was aware of what had to be done to recover his vision. At first, it was thought that he was trying something unorthodox in an attempt to gain advantage over his adversary. It soon became apparent that, if Rocky persisted in his tight turn, he could carry it all the way to the ground. The instructor pilot called to Rocky over the tactical frequency to get his attention — no answer. Suspecting, but not quite sure because he had never had a student pilot in such a situation, the instructor yelled over the radio instructing Rocky to let go of the stick, the only maneuver that would restore his sight and let him see to do whatever was then necessary to gain control of the aircraft.

It would be nice to report a happy resolution to Rocky's situation; but, watching in disbelief, the instructor saw Rocky strike the ground still in his tight turn — and most likely, still blacked out. That such an accident can happen, though very rare and decidedly unexpected, leaves many questions. Among them is the question why some pilots, however few, do not quite learn to believe their instructions and the hazards that ignoring them presents.

A Flight Worthy of a Distinguished Flying Cross

No book about Naval Aviation like this would be complete without the inclusion of a remarkable incident about a pilot and the venerable workhorse, the durable, reliable F-14 Tomcat, the rugged nature of its airframe, and its great F110 engines. Its ruggedness has allowed it to continue to perform even after serious damage, as in this incident worthy of the best in Naval Aviation and Naval Aviators. An attempt is made here to do justice to the heroic efforts of the pilot, then Lieutenant Commander Joe Edwards USN (later Commander) and the cooperation of the aircraft carrier personnel aboard the Nimitz Class nuclear-powered *USS Dwight D. Eisenhower* (CVN-69). This narrative is based on both an article in the Navy's flight safety magazine, Approach, of October, 1992, entitled "Topless at Twenty-nine Thousand Feet," which was written by Joe Edwards and his radar intercept officer Lieutenant Commander Scott Grundmeier, and also on written correspondence and oral communications with Commander Joe Edwards USN.

Although various details of this incident are well-known by many F-14 Tomcat pilots and supporting personnel, not all Naval Aviators are familiar with it; and the public at-large is generally unaware of it. As related in earlier sections, many Naval Aviators have faced and made decisions in challenging situations, some being relatively inconsequential, with others being quite serious, life-threatening, and even fatal. But anyone who has become familiar with the conditions challenging LCDR Joe Edwards USN in the Middle East in 1992 could hardly do other than agree that the recognition he received for his flying skills, his courage, and his perseverance in dealing with these conditions was well-deserved. A Naval Academy graduate of 1980, he was designated a Naval Aviator in 1982 specializing in Naval

Fighter Aircraft, particularly F-14 Tomcats, a specialization that was further enhanced by his subsequent test pilot experience. There is much more to know about the accomplishments, awards, and decorations of Joe Edwards — but later.

The F-14 Tomcat is a "Standard Fleet Fighter Aircraft," which came into service in 1972 and is expected to remain in service until the year 2007, yielding at that time both to the F-18 Hornet, now in service, and any new aircraft being developed. At about 65,000 pounds gross weight, it has two General Electric F110-GE-400 turbofan engines rated together at 40,000 pounds of thrust, which is increased to 60,000 pounds of thrust with afterburners. It has an internal fuel capacity of 16,000 pounds (about 2,666 gallons), increased to 18,000 pounds (about 3,000 gallons) by external tanks. It has variable swept-back wings which range from 20 degrees for takeoff and landing to 68 degrees in high speed and supersonic flight, and to a fully swept-back condition of 75 degrees for compact storage on the aircraft carrier deck. It has sophisticated avionics and multifunction flat screen (digital) displays referred to as "glass cockpit displays."

Emergencies of all kinds do not wait for a convenient time of day, for good weather conditions, satisfactory fuel states, or a convenient location. They can and do occur at any time and any place. But those conditions become important factors in what a pilot can do when emergencies come up. Under some adverse conditions, there is nothing the pilot can do except eject from the aircraft, a water landing (or ditching) in a jet aircraft being a very risky choice. But even in near ideal conditions of time of day, weather, fuel state, and location, other factors such as aircraft damage, pilot injury, equipment breakdown, or system malfunction (hydraulic system, etcetera) can create situations in which there is but little that can be done by the pilot — or that he can do only with great effort and skill, not to mention courage that requires a calm, intensely focused and persistent attention to the problem at hand. In LCDR Edward's incident, the emergency occurred in daylight hours and VFR (Visual Flight Rules) weather, free of clouds but hazy as the area near Bahrain frequently is. He and his Radar Intercept Officer (RIO) LCDR Scott Grundmeier USN were at twenty-nine thousand feet and thirty-five nautical miles from the *USS Eisenhower*, while Bahrain lay one hundred nautical miles in the opposite direction. Although conditions in the way of time of day, weather, fuel state, distance from ship and shore, etcetera were good, those would be the only good conditions Edwards would experience for some grueling time.

Those conditions represented the "good news." What happened next was the "bad news." Practicing radar air intercepts at 29,000 feet with another F-14 Tomcat flown by his "Skipper," Edwards thought he saw a distant airliner, the presence and monitoring of other aircraft always being important. To get a better look through the haze, he raised the tinted Plexiglas sun visor on his helmet. After seeing what he could and as he raised his hand to lower his visor, it happened! There was a loud bang of something crashing against the windscreen, and shattering and loosening the front, fixed Plexiglas panel extending out on both sides of the center; a torrent of Plexiglas shards blasted the pilot. Dislodged and missing completely were the clear, clam shell-like Plexiglas coverings of the pilot's movable cockpit canopy. Equally, or even more damaged than the aircraft, was Joe Edwards: his Plexiglas sun visor being up at the time of the crash, shards of glass had penetrated his face and struck both eyes, his right eye being hit so hard with shards that the iris had prolapsed. "(He) had several cuts on his face, glass in both eyes and quite a bit of pain in the right eye each time he blinked. Actually there were no large pieces of glass in his eye; the cornea had been lacerated by the debris and the iris had prolapsed, which caused it to cover his pupil. More accurately, the iris stretched until he had no pupil." Blood and pieces of flesh were spattered about; and since his visor had been raised, there was a red, sticky residue from Edward's face and eye which had splashed up against the inside surface of his helmet visor. Vision problems were serious, but they were only the beginning, because he suffered other serious physical problems as well: a broken right collarbone and a collapsed lung, all of which caused him, and would continue to cause him severe distress. Had the sun visor on his helmet been down, would his situation have been better or worse? Difficult to know. There was further damage to the aircraft which he noticed a bit later: his broken clavicle was caused by the radome or portions of it when it struck him and split the heavy chrome oxygen koch fitting at his right shoulder, damaging his oxygen fittings and severing his radio connections, or as Edwards put it, "leaving my oxygen hose and electrical connections… sitting in my lap. Could I hook them up? I need two hands."

Cut off from radio transmission and reception, he also was without his intercom. Adding to the confusion of the moment, were painful injuries, lack of oxygen, and air's rushing into the cockpit at an indicated airspeed of 350 knots (400 miles per hour), which was really much higher at 29,000 since his actual groundspeed as indicated on the mach meter was .88 percent or roughly 550 and 600 knots (or 625 to 685 mph). A lot of noise and its attendant confusion resulted. Still unsure of the source of the problem, he thought he must have struck some object. At the same time, RIO Grundmeier thought, "What the...? Never heard such a loud sound before. Something just brushed by the right side of my neck. Wind blast! Deafening roar. Can't hear anything. Max confusion. Adrenaline rush."

After the impact, and without fully knowing the extent of his problems, Edwards did not waste time, but immediately throttled back and put the aircraft into a tight four-G turn to get his nose down and spiral to a friendlier, more oxygen-rich altitude, descending at a rate close to 10,000 feet per minute or rushing down toward the ocean in a steep slope at approximately 120 miles per hour. In the process, he slowed his forward speed down to around 250 knots indicated. LCDR Edwards says he was not concerned about hypoxia (oxygen deprivation); and the noise did not bother him too much, since his actions seemed to be instinctive and focused on the more immediate. He placed his wheels lever in the down position and leveled off at 15,000 feet, relieved to find that his wheels went down and locked; it meant that, at least for the moment, he had hydraulic pressure. Would he also have hydraulic pressure later to get his flaps down? It is hard to imagine that any kind of a successful carrier landing would have been possible without wheels, and Edwards had ruled out any "no flaps" carrier landing. At first, he wondered what could have caused such damage but still left the aircraft flyable; but soon, he had pretty well reasoned that what had happened was that the radome, which covered his AN/AWG-9 Radar in the nose of the aircraft, had somehow come loose during a negative "G" maneuver, and the slipstream, with tremendous force, had peeled it up and back over the top hinge. On the way to the cockpit, the nose cone had broken the Angle of Attack (AOA) measuring device on the fuselage, which is invaluable for making carrier landings; it rendered it useless and the cockpit instrument indications meaningless. The nose cone also skewed the Pitot tube so that the airspeed readings became unusable, fluctuating to give a range of information likewise meaningless. Somewhere in all of this, Edwards glanced toward his RIO Grundmeier to determine his status, only to find him slumped forward, precluding any sure way of knowing his condition. What was left of the radome "passed down the spine of the Tomcat and through the twin tails. Debris had flown into the back cockpit, causing minor abrasions to the RIO's neck with pieces imbedding themselves in the headrest of the ejection seat."

With so many bits of Plexiglas having been propelled backward, Edwards felt sure that his engines had received some foreign object damage (FOD), saying, "I'll bet both of them took some hits. Gotta love those F110's (engines) though; they are still chugging along with no fluctuations in the engine instruments." Mindful of his condition and that of the aircraft, Edwards considered the desirability of flying into Bahrain, one hundred nautical miles away, comparing that choice with trying to get back aboard the *Eisenhower*. In any comparison of choices about Bahrain versus the *Eisenhower*, the desirable and understandable thought of going back aboard would be strong; but landing at Bahrain International Airport had impressive advantages. The runways at 10,000 to 12,000 feet in length allowed the luxury of a less precise approach compared to the very careful approach required for landing in the first few feet of a carrier's deck in order to engage one of only four arresting wires spaced fifty feet apart, a most difficult and demanding task for a twenty-five to thirty-five ton aircraft traveling at about 140 knots (about 160 miles per hour). Bahrain International's runway width also would provide a margin of safety, allowing the aircraft's approach to be somewhat to the left or the right of the runway's center line, enabling the aircraft to still land satisfactorily, whereas landing to one side of the carrier's center line would be hazardous. The approach speed for Bahrain could be higher — and more safe for a damaged, thus less aerodynamically clean aircraft — and still have the security of a long runway to slow down on after touchdown. While brakes would still be necessary to bring the aircraft to a stop, the loss of brakes would not likely result in either pilot injury or much further aircraft damage should the Tomcat run off the far end of the runway. Contrarily for the carrier, a higher approach speed might cause the plane to pass over the mere 150 feet of deck touch down space between the first and fourth arresting cables, necessitating a wave-off, more fuel consumption, another pass, and so on. Other problems can and do crop up in carrier landings. The pitching or rolling deck, while slight in Edwards' case, is not like a stable runway on solid ground. And finally, as efficient and prepared as carrier equipment and personnel are, there is always the possibility of delays caused by fouled decks and getting the flight deck crews ready for an immediate recovery. When you factor in all of Edwards' physical problems, the aircraft's problems, and then the uncertainty of the RIO's physical condition, the choice of Bahrain over the *Eisenhower* would, at first, seem all the more rational.

But naturally, Edwards had been thinking about all of these things and made a decision about a course of action reflected in his thoughts: "I need to choose between a bingo (return to an airfield) or CV (carrier) recovery. NORDO (no radio), blind right eye, injured shoulder, no canopy, no airspeed, no AOA. Not the best of situations to come aboard the boat. On the other hand, I probably have two FODed engines, an unconscious RIO. I'm NORDO with more than 120 miles of water to cross to land at a civilian field with unknown engine damage to the airplane, unknown medical facilities, and no way to get a visual inspection or communicate with anyone who joins us. Will the airplane make it that far? Fortunately, I have more than 12,000 pounds of fuel with a max trap (maximum weight of the aircraft's fuel load to land aboard ship) of 8,000 pounds. I could get a couple of looks at the deck and still bingo at low altitude with plenty of fuel." The complications with his cloudy vision, now a bit improved as the speed dropped down to an estimated 200 knots, and the serious problems associated with getting frequencies for navigation and radio transmission from a damaged communications panel, if he were to proceed to Bahrain, made choosing the *Eisenhower* seem a better choice — provided the Tomcat was flyable and enough of the aircraft's systems worked to make it possible. Joe Edwards was checking out his Tomcat, thinking, "The airplane is actually flying fairly well. Let's put some of that TPS (Test Pilot School) training to work... Three wheels down and locked — amazing. How about the flaps? Incredible! Fully down, no split flaps (one flap's position being different from the other). Things are looking up, but can I fly the ball (utilize the Fresnel Optical Landing System)? My vision is not that great. Let's see... left eye is pretty good, the right eye — uh, oh I can't see anything out of it. Surely, that can't be. Yep, no vision in right eye. Let's get a look in the mirror... great! A piece of glass protruding from my eye. I guess I can close out my log book for good after this flight. Can I recover this airplane with only one eye?" While depth perception is affected with loss of one eye, Edwards said that he did not sense any difference. Adding to the difficulties were the intercom problems, which precluded attempts between Edwards and Grundmeier to determine the overall condition of each other. But Grundmeier had tried to call Edwards: "Reb, how do you hear me? Nothing! Try the ship. 'Strike, Dakota 205, how do you read?' No one's talking today." Grundmeier had been leaning forward as if slumped over by injuries as he tried to limit the noise of the air rushing into their cockpit compartments. Only later did Edwards learn that it was Grundmeier's leaning forward for a better look at the radar scope just before the impact that allowed the Plexiglas shards to pass largely to the side and over his head.

With his wheels already down and having determined he would have flaps and speed brakes, Edwards would still need hydraulic pressure to get his variable-sweep wings to the desired swept-back angle of twenty degrees for landing. But even with a fully functioning hydraulic system, he still faced major problems. Flying aboard the aircraft carrier requires precise speed control and a precise angle of attack, or attitude, terms referring to nose position coordinated with engine power settings. Military pilots are imbued with the knowledge that maintaining the proper rate of descent for landing aboard ship, or even shore, is controlled by airspeed and the throttle, provided the nose position is properly established. Such a combination is required for following the electronic glide slope down through the clouds during instrument landings at air fields and also on aircraft carriers, although a more exact system is required for the strict demands of flying aboard ship. But in Edwards' situation, it was knowing how to establish the required attitude without airspeed indications that would be his severe challenge.

So far and as you would expect from his background, Edwards had taken things logically and in the right sequence. When he was not completely sure what he had struck or what had struck them, he had dropped to an altitude where oxygen was not necessary and where he could start to assess the damage and whether they could stay with the Tomcat or whether they would have to bail out. Checking out what was still available to them in the way of operable equipment and systems was the next step. Would the air-

More Surprises Bring More Challenges 331

craft be flyable in the mode required for carrier landings or would a bingo be a better choice? Having determined that it would be flyable, despite the lack of airspeed and angle of attack information, he reasoned that he could use other means for approximations of that information. Further, he would try a couple of practice approaches to get some idea of what he would have to contend with. So, he was ready to try to set up the Tomcat in the "approach mode" and start a trial run. But first, he wanted to try to inform the ship of his condition and indicate that he would make a trial run.

Assuming the aircraft was airworthy enough to make an attempt to land on the carrier, what he, or any other Naval Aviator, faced in an attempt to fly back aboard the carrier may be reliably described by Peter Garrison in an article in *Aviation Week and Space Technology* dated May, 1995. He provides some perspective and appreciation of what any healthy Naval Aviator flying an undamaged aircraft faces as he makes his approach: "The pilot must control speed, attitude, and position with microscopic vigilance. At the aft end of the flight deck — 'at the ramp' in the carrier pilot's jargon — the window through which the arresting hook must pass to catch the target third wire (of only four) is only two feet high. An aircraft weighing twenty or thirty tons (the weight of 16 to 24 mid-sized automobiles) and spanning twenty yards must reach that narrow slot at precisely the right rate of descent and attitude. The hook dangling from the airplane's tail is supposed to clear the ramp by at least ten feet; but the ramp itself, in heavy seas, may be rising and falling fifteen feet and sashaying side to side in a figure eight as well. And all the elements — speed, attitude, height — are linked. A change in one causes changes in the others. Carrier landings resemble those games in which you try to roll one ball bearing into its well without allowing others to roll out of theirs. The lightest penalty for imprecision is humiliation. The gravest is sudden and violent death." (As mentioned earlier, "ramp strikes" have occurred in all types of aircraft, including F-14 Tomcats; but they have been very, very few in number compared to the huge number of successful approaches.) As challenging as Garrison's description sounds, he described a healthy pilot and a healthy plane. Edward's situation had neither. And the pilot had no radio communications which would allow him to receive help, advice, and encouragement from the landing signal officer should it be necessary. Beyond these problems, lay Edward's ever-present awareness that the information he needed most, airspeed and angle of attack information, was not available to him.

Edwards tried to set up a sample configuration thinking, "The jet's still flying well. I can't trim it up (to fly) hands off, but controllability is good. Let's slow up to 'on speed' (carrier approach speed)... great: the airspeed indicator is bouncing around between 30 and 160 KIAS (knots of indicated airspeed). I guess there was damage to that probe too. Well, I know how much power we need to maintain 'on speed.' I'll set 4,200 pounds per side and see what that gives us. On second thought, make it 4,800 pounds per side to give us a little gravy." A little later Edwards thought, "I'm near 'on speed' now. Longitudinal trim (fore and aft trim) is about right, power is set and it flies pretty nice. Let's start a descent. We're about ten miles out. Can I make large enough control inputs with this shoulder to wave-off? Yep."

Despite having considered the possibility of ejecting as one possibility in a list of things that might be necessary, and despite his physical condition and that of the aircraft, Edwards says that he did not seriously consider ejecting, though it still remained an option. Should he decide to eject, one can only imagine the additional pain and possible further damage to his shoulder from the force of a rocket-propelled ejection from the aircraft and the opening shock of the parachute which would blossom instantly. Would he have the strength to free himself fast enough from the parachute harness as he was about to enter the ocean — and, once in the water, could he work himself free should he become entangled in his chute? He would likewise have had to consider the harsh effect of salt water in his shard-impacted face and eyes. No doubt, these considerations made him lean toward a landing either aboard ship or on land as much more preferable to bailing out. Besides, though RIO Grundmeier seemed more alert, Edwards, not having any contact with him, really did not know his condition.

Normally in daylight, Naval aircraft approach the carrier along the starboard side at the specified altitude of 800 feet and, when slightly ahead of the carrier, make a 180 degree descending turn downwind to 600 feet, slowing down from about 300 knots to about 150. The prescribed distance abeam of about one nautical mile is maintained so that a comfortable "approach turn" toward the carrier can be initiated when the aircraft is about even with the ship's stern. It is early in this approach turn that the pilot starts to set up his airspeed and attitude or angle of attack. Since the aircraft's speed is slower and closer to the stalling speed and the nose is higher, the pilot must be careful and precise to insure that stalling and losing the lift necessary to keep it airborne not occur. Should the F-14 stall at such a low altitude, there would be no way to keep from crashing into the ocean. The procedure for Joe Edwards included the possible hazard of unknown and further aircraft and engine damage and required that he consider the effects of the now blunt-nosed aircraft and the resulting drag and other aerodynamic effects of the missing radome and the shattered windscreen. But before he could attempt an approach to the carrier, he thought, "Let's get to a lower altitude and see if this thing is flyable in the approach configuration before we decide. We may have to eject alongside the ship. Better tighten up those lap restraints with my left hand." — But would it be possible to set up the aircraft in the approach configuration, all without the most important and helpful indicators carrier landings in jet aircraft require?

With no communication from Grundmeier, he felt that somehow he had to relay his condition to the *Eisenhower*: "We've got to let the ship know we're out here. A dirty pass (wheels and flaps down, wrong side of ship, etcetera) at 300 feet up the left side of the ship (whose flight deck is 60 feet above the water) should do it. Let's just hope they see my canopy glass is missing. Rock the wings and cycle the throttles, try to tell them we're NORDO. Turn downwind. I can't see anyone on the platform. Dump a little fuel, recheck that landing checklist." In the meantime, Grundmeier continued to try to contact the *Eisenhower*. With the roar of rushing air diminished at the lower airspeeds, and by squeezing the sides of his helmet, Grundmeier called, "Paddles (The landing signal officer), Dakota 205," and was relieved to hear, "This is paddles. Go ahead 205." "The ship knew something was wrong with Dakota 205. The aircraft was squawking emergency IFF (a transponder which produced a coded blip on the ship's radar) and flight deck personnel saw the plane make a low fly-by. The ship immediately began to clear the deck and prepare for an emergency landing. No one on deck noticed that the canopy glass was missing, but they did see that the radome was gone. They knew the aircraft was in trouble and wanted to land ASAP. LSO's were called to the platform, and aircraft were cleared from the landing area." Grundmeier explained the situation to the LSO's and called, "Tower, 205, request medical personnel to stand by on the flight deck." The Tower answered, "Roger."

The trial approach idea seemed appropriate, provided he could see and stand the pain of the air coming into his face and into his damaged eyes, wondering that when it was, "Time for a peek up forward, I don't get my face ripped off in the windblast." He made two practice approaches using the daytime approach turn method. It was in these passes that he discovered airspeed problems were real and persistent, while what he really needed was a steady reliable reading. RIO Grundmeier was thinking, "We flew two invaluable approaches as the ship turned into the wind, allowing us to fine tune techniques used in all those FCLP (Field Carrier Landing Practice) emergency practice periods." In these trial approaches, Edwards knew that he could see the aircraft carrier deck and lens better by poking his head out into the slipstream. But it was decided that a long straight-in approach often used at night would be better. This was also often the type of approach made in bad weather. It required that a straight-in approach be made from a position some point aft of the carrier.

Now, having established what he had to work with, he was ready to start the actual carrier approach. Having definitely decided to land aboard the aircraft carrier, and optimistically assuming that the aircraft's condition was not likely to change, Edwards actually started his approach from astern at about three miles distance and 1200 feet of altitude, an altitude and distance which would place him in the right position to approach and soon engage the Optical Lens Landing system. The OLS would be the means of effecting the precise approach described by Peter Garrison mentioned earlier. From the "right position," Edwards would fly the orange light or "meatball" emanating from the OLS, which, when lined up with a row of green lights also emanating from the OLS, would indicate the aircraft is right on the glide path course to the carrier. If the "meatball" appeared above the green row of lights, Edwards would know he was too high; if below, he would know he was too low and would have to make a correction. Absolutely critical in all of this is the attitude and airspeed; ignoring them or being careless with them is disastrous. As he started, and lacking information he would normally have had to assist him, he started to employ other less precise means of approximating that information. Nevertheless, he thought, "Longitudinal trim (nose position maintained by the elevator or adjustable horizontal tail surface) is about right, power is set and it flies pretty nicely. Let's start a descent." When he caught sight of the "meatball," which was largely obscured by the shattered windscreen Plexiglas, he had to poke his head into the slipstream continuously to keep the crucial "ball" in sight — he had no choice, even as his approach speed was about 140 knots (close to 160 miles per hour). He could use information from the Instrument Landing System ILS as a cross reference up to a certain point in his approach; but of course, nearing the carrier, he just had to rely on the OLS to maintain the precise attitude, airspeed and rate of descent to land aboard. Since neither Edwards nor Grundmeier had airspeed or AOA information, Edwards had to depend on setting and adjusting approximate engine power, trimming the aircraft longitudinally with the movable horizontal stabilizer of the tail (which has replaced trim tabs used in earlier aircraft types,) and estimating control stick forces in level flight to set himself up as best he could to obtain the desired 'on-speed' condition. To get an estimate of the airspeed required to descend on the glideslope, he set 3,400 pounds of fuel flow (an indication of engine power) per side, which was approximately 100 pounds per side greater than required at max trap weight in the aircraft's current configuration. Though Edwards had determined that, if necessary, he could take a wave-off and make another approach, he really wanted to get aboard on the first time around. A second attempt would hardly be easier. Yet his extensive background and his experience as a test pilot sharpened his ability to set things up properly the first time so that only small adjustments to power and trim would be necessary. Knowing approximately what power settings were required to land at maximum trap weight (8,000 pounds of fuel), he determined the extra 100 pounds per engine he had decided upon would offset the drag caused by the missing radome and the canopy damage.

Considering all that had happened to this point, it would be easy to conclude that, once being able to spot the orange light from the Optical Landing System, Edwards had it made and would make a smooth approach to the carrier deck, easily bringing his ordeal to a satisfactory conclusion. But that would give the false impression that everything was easy and almost automatic once the pilot caught sight of the orange light. After all, the OLS does not take over; on the contrary, it demands of the pilot skill and precision which require greater concentration and persistence, any carelessness exacting a price to be paid. All Naval Aviator carrier pilots recognize the tenseness of getting themselves into the critical position, from which they will receive the "cut signal" to land aboard. Given Edwards' airspeed of 140 knots and the aircraft carrier's heading into the wind at about thirty knots, the aircraft's speed relative to the aircraft carrier was about 110 knots. This relative speed meant that, if Edwards used the OLS exclusively from about three miles out, he had to ride the "meatball" down for a period of nearly two minutes — enough to set up the attitude and power setting which would control the rate of descent, and also his azimuth or heading, provided his skills allowed him to avoid making big changes which would likely result in overcompensation and the resultant wave-off.

How many experienced carrier pilots would attempt to do what Edwards was about to do cannot be known. All would know the correct engine power settings, attitude, and approach speed for an undamaged aircraft. All would know how to profit from the angle of attack indicator to remain safe from stalling. And all would know how to adjust power, and perhaps nose position, to stay on the glide slope of the Optical Landing System — even if required to make multiple adjustments. But, with all of the unknowns plus one-eyed and blurred vision and a smashed, vision-obstructed windscreen, all of which would have to be contended with while enduring the pain and discomfort of injuries, certainly many would opt for a less challenging and dangerous landing at Bahrain.

It seems so simple to say, "Not surprisingly, Edward's estimate of engine power needed in his damaged aircraft and his use of the horizontal stabilizer (which in an undamaged aircraft ordinarily allows a nearly hands-off elevator setting and a somewhat stable nose position) to maintain altitude resulted in a pretty close approximation of that needed. Knowing that he would ride the "meatball for over a minute and a half, Joe Edwards was able to establish and carefully maintain this difficult relationship of power and longitudinal attitude with minor adjustments, which kept this skillful pilot on the glideslope — maintaining, within limits, attitude, descent, speed, and azimuth — despite all the challenges presented — to bring the Tomcat right down to the deck and the desired Number Three arresting cable. It was a finish to a challenge that very easily could have had a disastrous ending!" So simply stated, and for the uninitiated, there may be a tendency for some to underestimate the use of acquired skills and the courage required to accept, complete, and face the challenge! That would be understandable; but for experienced Naval Aviators, no more need be said.

Safely on the carrier deck, Joe Edwards faced one more challenge — trying to reach the ladder release handle which would enable him to exit the aircraft and climb down to the deck — a procedure usually requiring the use of the F-14 pilot's right hand. But this too he overcame by awkwardly using his left hand.

No one out there believes landing on aircraft carriers is easy, but constant training and good equipment make a big difference. The improvements in landing devices like the latest "Improved Fresnel Lens Optical Landing System" work hand-in-hand with field practice. It has been said that a Landing Signal Officer can tell an approaching pilot's angle of attack better than the pilot. No doubt their experience helps them determine the aircraft's attitude fairly well; but people remembering the old paddles system and the kinds of incidents which occurred would not necessarily agree, nor does this author. Regardless, you can be sure that the field practice, the Optical Landing System, and Joe Edwards' skills enabled him and Scott Grundmeier to get safely "home," at the same time saving a very expensive, damaged but repairable, aircraft for further use.

To appreciate what the Optical Landing System (OLS) does, a comparison with other runway approach systems which are used to help the pilot descend through bad weather is helpful. There has been an evolution of instrument landing systems: the old "Radio Range System," from which came the term "flying on the beam;" the "Instrument Landing System (ILS);" the "Ground Controlled Approach System (GCA);" and currently the Omni/DME or its military cousin "TACAN." All but the ancient "Radio Range System" are still in use. These systems would normally be utilized for relatively longer periods of time than the OLS: for example, five or more minutes during a letdown from high altitude in bad weather compared to the 90 or so seconds during the OLS approach. All of these systems are electronic in nature, but with significant differences. And while all could be helpful in bringing the pilot down to the near vicinity of the aircraft carrier, none would produce the precision required for safe carrier landings. Rather, they should be looked upon as complementary systems, which at sea can bring the pilot down through bad weather to a point where he can pick up the Optical Landing System. The OLS is, on the other hand, quite different in that it sends out a beam of light at a set angle, which, through a system of lights and Fresnel lenses, provides the requisite information from which the pilot can control azimuth, attitude, and rate of descent. The full name of the current OLS is the "Fresnel Lens Optical Landing System," after the French physicist Augustin Jean Fresnel (1788-1827), who developed the system of lenses and mirrors often used in lighthouse beams. Unlike the other systems which produce an electronic glideslope, it sends out a beam of light providing a visual glideslope that, if properly and carefully flown by the pilot, will result in the aircraft's approaching the aircraft carrier's deck precisely through that very small window necessary for a safe arrested landing on a very limited piece of the carrier's deck. It is that precision that makes this system so much better for carrier approaches than any other system yet devised. But that precision demands very careful concentration and skillful flying. As emphasized, critical to the success of OLS approaches is the aircraft's speed, which the pilot can only achieve by having the combination of the correct attitude or nose position (or angle of attack) and the precise power setting to maintain that attitude — and hence airspeed. The process involved in landing aboard an aircraft carrier is the most demanding, therefore most tense, maneuver in Naval Aviation. The stress placed upon the pilot is severe and has been described in various ways to indicate that level of concern. Thomas Wolfe in his book *TheRightStuff* uses hyperbole in likening it "to heaving a 50,000 pound brick into a skillet." And yet, however stressful day landings may be, nighttime makes them more so.

But whether the glideslope is electronic or visual, there is nothing automatic about descending on it. With the Optical Landing System's being more precise and demanding, it is rare that the pilot — or the occasion — exists when the pilot can make a correction to his attitude, power, or speed during the approach and then depend on never making corresponding adjustments. Any correction he makes calls for another adjustment. Any change in heading or altitude will require a change in power setting to correct; then, back on course, or heading, that power setting will have to be adjusted again. And even if the pilot establishes his position on the glide slope, he must maintain it through changes in wind force and direction and also through turbulence. The bigger the deviations from the glideslope or heading, the more extreme the corrections necessary and the more likely the pilot will have to make another pass, with the result that his jet engines will "guzzle" fuel frighteningly fast and place additional stress on the pilot. It was the desirable settings which Edwards maintained and the undesirable things which he avoided that made his approach so noteworthy. A working Angle of Attack indicator, a very simple vertical dial in the cockpit would have indicated his proximity to stalling; and a properly functioning airspeed indicator to accompany the AOA indicator would have made setting up and maintaining the approach so much more easy — without them it's a very difficult challenge.

Some people have described the OLS light signals and system as a funnel, with a wide opening at the beginning of the approach, which narrows down as the pilot gets close into the ship. Such a description might imply that, once the pilot is in the funnel, the approach is automatic and he is being funneled easily toward the carrier deck. A more preferable, yet still inexact, way of describing it is as a very narrow cone rather than a funnel, because the pilot starts out at a more exact position than the wide opening of the funnel, one that would get him more precisely on course and glideslope at the beginning of his approach, narrowing down the size of adjustments necessary later. For example, the orange "meatball" emanating from the OLS is a beam of light only one half degree in width; so, even at a distance of one mile astern the carrier, the beam is only twenty-three feet in diameter. If the pilot is in the middle of the beam at one mile distant, he need only vary twelve feet before he is off the glideslope. However, using the information provided by the "Improved Fresnel Lens Optical Landing System," there is a three quarters of a degree area both above and below the yellow light in which the pilot will also receive a green light if above the glide slope, and a red light if below. Greater variations will produce flashing green and red lights, requiring more radical maneuvers to correct and get back to the "meatball" and requiring the likelihood of a wave-off and another approach.

That Joe Edwards could handle the stringencies of the OLS in his physical condition and in a battered aircraft, all the while in pain and much impaired vision is truly impressive — that he did it so well is even more so.

After Edwards' exceptional job in "trapping" the third wire aboard *Eisenhower*, he was treated medically, at one time having nine IV's attached to him. Later, he was flown by helicopter to the Bahrain International Hospital where they stitched up his eye, took care of his clavicle, treated his collapsed lung and kept him for ten days. While Joe Edwards thought his flying days might be over, in four and a half months his injuries had healed satisfactorily and his rehabilitation had been completed, placing him once more in an "UP" flying status. So well recognized and admired was his performance that he was awarded the Distinguished Flying Cross by the Secretary of the Navy, one of only very few awarded in peacetime.

Epilogue

Throughout its history, Naval Aviation has always had an outstanding cadre of aviators who, by example, have led the way, taking on difficult challenges, demonstrating how the nearly impossible can be done, and inspiring others. There are always some Naval Aviators who continue to emerge and belong in that select group. Of more recent times are several shown to belong by virtue of their significant achievements. Two of that cadre are mentioned in this book, and a third is added, but there are many, many more who could be included. It is always a problem when names are mentioned, a problem that others, equally as important, are omitted. But trying to include all would be quite long and in this undertaking impossible. Their omission should not be taken as a slight in any way; but rather it is necessary for readers to know just a few names to realize the kind of Aviators in this group. Surely, knowledgeable Naval Aviators know many others who could be included and wonder why these three have been mentioned to the exclusion of others whom they think equally worthy. On the other hand, those same Naval Aviators could have nothing but admiration for those selected here to represent the group.

Of the two mentioned in this book, here is Captain Dale Snodgrass USN who, according to "Tomcat Notes," a publication of the Tomcat Association, is listed as the Navy's top Tomcat (F-14) pilot in terms of experience and hours flown (4600 hours in the F-14 Tomcat). His total number of squadron tours of duty, his total flight time, and his number of carrier landings are truly impressive. "Fighter Pilot of the Year" in 1985, Grumann Aerospace's "Topcat of the Year" in 1986, and a Top Gun graduate, he became the Fighter Wing Commander of the Atlantic Fleet. Among his many, many awards, are the Legion of Merit (3) for superior performance in positions of great responsibility and the Meritorious Service Medal (2) for exceptional service in a position of senior leadership. Much, much more could be said. The second is commander Joe Edwards USN, whose Distinguished Flying Cross is described in the last episode of this book. He was "Fighter Squadron Pilot of the Year" in 1984, 1985, 1990, 1991, 1992, and "Carrier Air Wing Seven Pilot of the Year" in 1985, 1900, and 1991. He earned a BS in Aerospace Engineering from the Naval Academy in 1980 and was a graduate of the U.S. Naval Test Pilot School, later becoming the Project Flight Test Officer for the latest Tomcat, the F-14D, and an astronaut for the next-to-last Mir Station Mission. He holds an MS in Aviation Systems from the University of Tennessee. Currently, he is president and CEO of the National Science Center, Inc. in Augusta, Georgia. Likewise, much, much more could be said about Joe Edwards.

Exceptionally worthy beyond doubt, but not mentioned in this book so far, is Commander Charlie Ernest USN. So capable and a proficient pilot and leader, CDR Ernest lost his life in a fluky accident caused by mechanical failure. Former Navy Secretary John F. Lehman in his book *Commander of the Seas* refers admirably to this veteran of 350 combat missions over Vietnam in his description of the accident: "As his A-6 Intruder, loaded with bombs, accelerated down the catapult track, the cathode-ray display, directly in front of the stick, on his instrument panel broke loose because of defective bolts and landed in his lap, jamming the stick back. The jet did an immediate hammerhead stall right off the end of the catapult track, and with the huge box in his lap, Charlie was unable to eject. The bombardier was able to eject the split second before the aircraft hit the water. Another great warrior, Admiral Jack Christiansen, was watching from the bridge, and it was he who presented posthumous awards to Charlie's son, Brad, months later. He accepted his father's three Silver Stars, seven Distinguished Flying Crosses, the Bronze Star, thirty-two Air Medals and various other ribbons." Again, of Commander Charlie Ernest, much, much more could be said. It is ironic that, although this accident was operational in nature and occurred in a combat zone, it was not, strictly speaking, combat-related. Rather, it took place in what was supposed to be a more routine phase of operations, emphasizing the overall, challenging environment of Naval Aviation.

In mentioning this cadre of Naval Aviators, there is no intention to diminish the achievements of any other Naval Aviators, aircrew, junior officers, or senior officers, many of whom could be equally cited here. But, however many be cited, they still could not do everything Naval Aviation requires, relying therefore on the main body of lesser-known, but very capable, Naval Aviators and aircrew, who constitute a very effective force that carries out the bulk of naval Aviation's responsibilities, flying round-the-clock missions in all sorts of weather, often far out to sea and reach of land, and many times in hostile environments.

There are many in Naval Aviation who would fit the description used by James Michener in his book *The Bridges of Toko Ri* when he has Admiral George Tarrant say of LT Harry Brubaker, who was called to active duty, and killed during the Korean War, "Why is America lucky enough to have such men?... Where did we get such men?" It does not seem inappropriate to ask, "Where indeed?"